Innovation Systems for Development

Innovation Systems for Development

Making Research and Innovation in Developing Countries Matter

Edited by

Bo Göransson

Research Policy Group, School of Economics and Management, Lund University, Sweden

Claes Brundenius

Research Policy Group, School of Economics and Management, Lund University, Sweden

Carlos Aguirre-Bastos

Institute for Social and Economic Research, Catholic University of Bolivia

EE Edward Elgar PUBLISHING

Cheltenham, UK • Northampton, MA, USA

Published by
Edward Elgar Publishing Limited
The Lypiatts
15 Lansdown Road
Cheltenham
Glos GL50 2JA
UK

Edward Elgar Publishing, Inc.
William Pratt House
9 Dewey Court
Northampton
Massachusetts 01060
USA

A catalogue record for this book
is available from the British Library

Library of Congress Control Number: 2016942162

This book is available electronically in the Elgaronline
Economics subject collection
DOI 10.4337/9781783473830

MIX
Paper from
responsible sources
FSC® C013056

ISBN 978 1 78347 382 3 (cased)
ISBN 978 1 78347 383 0 (eBook)

Typeset by Servis Filmsetting Ltd, Stockport, Cheshire
Printed and bound in Great Britain by TJ International Ltd, Padstow

Contents

Contributors

Carlos Aguirre-Bastos, Institute for Social and Economic Research, Catholic University of Bolivia.

Claes Brundenius, Research Policy Group, School of Economics and Management, Lund University, Sweden, and UNU-MERIT, Maastricht, the Netherlands.

Tran Ngoc Ca, National Council for S&T Policy (NCSTP), Vietnam.

Raúl Rubín de Célis Cedro, Institute for Social and Economic Research, Catholic University of Bolivia.

Sérgio Chicumbe, National Institute of Health, Ministry of Health (MISAU), Mozambique.

Lanta Daniel, Science, Technology and Innovation Policy Research Organization (STIPRO), Tanzania.

Maximiano Dgedge, Ministry of Science and Technology, Higher and Technical Education (MCTESTP), Mozambique.

Bitrina Diyamett, Science, Technology and Innovation Policy Research Organization (STIPRO), Tanzania.

Bo Göransson, Research Policy Group, School of Economics and Management, Lund University, Sweden.

Justine Liberio, Science, Technology and Innovation Policy Research Organization (STIPRO), Tanzania.

Javier Aliaga Lordeman, Institute for Social and Economic Research, Catholic University of Bolivia.

Heric Thomas, Science, Technology and Innovation Policy Research Organization (STIPRO), Tanzania.

Ignacio Garrón Védia, Institute for Social and Economic Research, Catholic University of Bolivia.

Acknowledgements

In the making of this book, the editors are greatly indebted to a large number of researchers, policy makers and representatives of the business community and civil society taking part in national workshops or other research activities. Their insights and ideas communicated through national dialogues and surveys in four countries – Bolivia, Mozambique, Tanzania and Vietnam – have been invaluable for the writing of this book. In particular, we would like to collectively thank members of the UniDev research network for critical and constructive comments on early drafts of the chapters in this book.

For financial support we thank the Swedish International Development Cooperation Agency (Sida), which provided funding for the research project on which this book is based.

1. Science, technology and innovation for whom?

Bo Göransson

1.1 INTRODUCTION

With the rise of organized scientific research, we have grown accustomed to an unceasing delivery of new scientific results and technical improvements that resolve even seemingly unsolvable problems. The share of humanity that reaps the fruits of this tremendous technological progress, however, has been and continues to be unevenly distributed around the globe as research efforts are typically directed towards solving problems facing an urbanized Western population. In the 1990s, the Global Forum for Health Research estimated that only 10 per cent of the world research effort is devoted to meeting the health needs of 90 per cent of the population. Although progress has been made since then (GFHR, 2004), this bias towards population groups with higher purchasing power persists and is not confined to the health area.

Perhaps more surprising is that research done by researchers in developing countries also has a similar 10/90 gap. Knowledge exists, but there do not seem to be enough interest or incentives in the innovation system to solve urgent problems of poor communities. This begs the following questions: Who sets the research agendas? Are they mostly decisions of the researchers themselves? Are they the priorities of the funding agencies? Is the research agenda the demand of those who are able to pay for research, for instance, pharmaceutical companies, or is it user-driven and inclusive, set by those most in need of the research results? And last, but not least, how are the scientific results used to inform policy makers?

This book addresses these questions by examining how science-based research and innovation is designed, implemented and applied in developing countries in support of development and poverty alleviation. The relatively scarce resources devoted to higher education and research in these countries are in many cases not directed squarely towards addressing pressing developmental issues and poverty alleviation. Even in cases of successful local research and innovation outcomes, appropriate framework conditions for supporting the application of the innovation may be lacking.

In the following chapters we provide examples of how the research agenda is set by the major actors involved in the national innovation system in four developing countries – Bolivia, Vietnam, Tanzania and Mozambique – and discuss how the results of the knowledge creation process can be better integrated into a national development strategy. In doing so, we focus on the expectations that are increasingly linked not only to the roles of science, technology and innovation (STI) in the development process but also to its results. There is an increasing concern among governments, policy makers, donors and other stakeholders that there is a mismatch between research and development (R&D) input and its results. Much money might be spent on R&D facilities without producing a dividend that is significant, or perhaps cannot even be measured. For developed countries it can be frustrating to not get the sought-after mileage of the investment. For developing countries the effects may be more sinister and, in the worst case, prevent development.

Bridging the gap between the production of knowledge and the application of knowledge in support of development is thus vital for the efficiency of the innovation system and its ability to address problems in development. Throughout this book we shall use the concept of 'Closing the Loop'[1] as a euphemism for efforts to align the output of the research community with the intentions of the policy makers and, thus, make research matter more to developmental goals. By Closing the Loop, the relevance and utilization of research outputs increases, which in turn enhances the influence of researchers and institutions in the development process. Likewise, the prospects for evidence-based decision making and informed policy formulation increases, thereby supporting a process of mutual reinforcement between the research community and the policy-making sphere.

The book thus contributes to a better understanding of how the process of knowledge creation and innovation in developing countries can be made more conducive to the development efforts pursued. More specifically, it presents evidence on the emergence of innovation systems in developing countries, including an analysis of the converging or diverging expectations of the main actors in the innovation system in the case countries. Further, it identifies the main impediments to enhancing the relevance and quality of research outputs for the productive sector, and offers suggestions on how the process of applying research outputs and innovations in support of developmental goals can be enhanced.

1.2 BACKGROUND AND RATIONALE

Addressing the impact of limited resources has been a preoccupation for decision makers and policy for a long time. Since the dawn of the economist profession, the idea of a self-regulating market system in which the use of productive forces is maximized by itself has had numerous and vocal proponents among economic thinkers. But, as shown by critical voices from different shades of the political spectrum, most recently by Thomas Piketty (2014), an unrestrained economy tends to lead to economically and socially suboptimal and inequitable outcomes. The prescribed antidote by legislators has been the introduction of varying degrees of redistributional mechanisms into the economy to alleviate undesirable effects. This calls attention to the role of public policy. History tells us that policy indeed matters and examples abound of successful public policies, be they aimed at fostering up-and-coming, fast-growing firms or creating competitive advantages of entire regions or nations. The phenomenal economic growth and technological catching up of some Southeast Asian countries in the 1970s and 1980s and, more recently, China are a case in point. In such cases, policy makers have clearly been successful in 'picking the winners' rather than leaving it to the market to 'create through destruction' in Schumpeterian terms. Whether or not such examples of successful public policy are the result of informed decision making and prudent interventions by an entrepreneurial State or just instances of pure luck is a bone of contention among economic scholars. Disastrous examples of failed industrialization policies can indeed also be found, such as the Great Leap Forward in China in the late 1950s or the blanket import-substitution policies of India up to the 1990s, which point to the risks involved in excessive or damaging State regulation. Moreover, the failed interventionist policies in the West in the 1970s in attempting to breathe new life into sunset industries implies that the case can be made for an exceptional and very specified role for State intervention rather than to mitigate private industry shortcomings. Or as Mariana Mazzucato puts it in her advocacy of the entrepreneurial State: '[W]hile the State needs to take risks, it should not be simply absorbing . . . . the risk of the private sector, but taking the kind of risks that the private sector is not willing to take' (Mazzucato, 2013, p. 194). We can extract two essential components for successful governance – an effective policy that is properly designed to achieve what it sets out to do and which rests on relevant knowledge capturing the state of the art of evidence on the particular problem it is meant to address. The process of turning such knowledge into policy is at the core of this book.

The area of STI is one of the few areas where there is almost total consensus for governmental intervention (be it for slightly differing reasons) for guiding scientific efforts today. Depending on the view, without government support the production of new knowledge will either be quantitatively suboptimal as a result of a market failure, fail to materialize in strategic basic research areas or even result in the 'wrong kind' of technology from a societal perspective.

The impact of STI on growth and development is increasingly being recognized. In the Organisation for Economic Co-operation and Development (OECD) countries this has been part of the agenda since the 1960s. It is more recently that STI policies are receiving the same attention in developing countries although a few pioneers such as Michael Moravcik had already in the 1960s argued that a broad range of pressing problems in developing countries could not be solved without an indigenous capacity for technological development (Moravcik, 1966). Today, however, such policies are seen as a *sine qua non* also in the South. We can see a growing interest in and need for measuring R&D input (R&D human resources, R&D expenditures and so on) as well as R&D output indicators (for example, patents, scientific publications and so on). Weighing these two aspects against each other one could get an approximation of the efficiency of 'the innovation system'. This has led to discoveries of so-called paradoxes, or mismatches. It started with 'the European paradox' when Dosi et al. (2005) claimed that while European Union (EU) countries were doing seemingly well in scientific output, this was not reflected in innovative performance. A year later attention was drawn to the 'the Swedish paradox' (Ejermo and Kander, 2006; Jacobsson and Granberg, 2006), with a discussion of the reasons why Sweden, topping the league of R&D expenditures as share of gross domestic product (GDP), apparently did not do so well in terms of innovative capacity.

These controversies that started as typical developed country problems/debates are now followed by similar discussions in developing countries. While there has been a growing concern about the need for R&D and innovative capacity building in the South, little attention has been given to the output, and the efficiency (and even relevance) of the R&D systems in these countries. It is in this context that innovation systems in developing countries are being addressed (see Lundvall et al., 2009) and it is also in this context that this book is placed. The relatively limited resources that developing countries devote to R&D make the efficiency and internal functioning of the innovation system an even more pressing issue. Of the four case countries in this book, Mozambique, Tanzania and Vietnam spend around 0.5 per cent of their GDP on R&D with Bolivia trailing with 0.11–0.25 per cent. This can be compared to the 2.9 per cent of GDP devoted to R&D in the USA, the OECD average of 2.4 per cent and to top spending Finland and Sweden investing 3.9 per cent and 3.4 per cent, respectively, according to OECD statistics for 2010.

1.3 CLOSING THE LOOP INCREASES THE RELEVANCE AND UTILIZATION OF RESEARCH OUTPUTS

Given the limited resources in developing countries as well as a limited policy formulation capacity, how can the scientific community in developing countries better support the developmental goals formulated in national development plans? Evidence on such a knowledge-to-policy process indicates that there is a sometimes hard-to-bridge chasm between the research community and the policy-making community and that there exists a degree of mutual distrust between the two worlds. The reasons appear to be quite universal. In studies from different countries and levels of development on the use of academic research by policy makers, a number of common reasons for not relying on academic research is provided. First, the relevance of academic research may be perceived as not having a direct bearing on the problem at hand, either because of a level of certainty too low to make it relevant (Jefferys et al., 2007), quality that is too low (IDRC, 2002) or due to researchers tending to overanalyse problems and not able to produce research results in time (Hy et al., 1995). Second, research is often regarded with suspicion by policy makers who may see it as theoretical exercises with little or no practical applicability in the real world. Or, put slightly differently, '[S]ignificant elements [of policy makers] seem to regard "research" as the opposite of "action" rather than the opposite of "ignorance"' (Surr et al., 2002, p. 1). Third, it is important to point out that policy makers have a host of other information sources and political realities to adhere to, making research only one, albeit vital, of several potentially influencing factors in the policy-making process.

For the research community, the most often mentioned factor hampering the provision of policy-relevant information to policy makers is the lack of institutional incentives (Jefferys et al., 2007; SPP, 2009). Academia functions with its own internal logic with an incentive structure that predominantly rewards publication in academic journals and can act as a deterrent to researchers' involvement in advisory or other extra-curricular activities. Researchers and policy makers rarely move in the same circles, pointing to the lack of established channels through which research results can be presented (Court and Young, 2006). Additionally, researchers in general are not trained or experienced in communicating with policy makers (Carden, 2004; Start and Hovland, 2004).

Bringing these and other issues together in an analytical framework for bridging research and policy, Court and Young (2006) identify four overlapping areas where a number of factors influences the likelihood

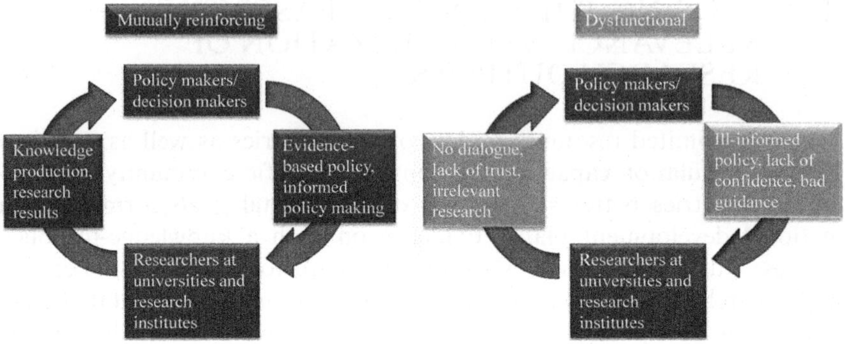

Source: Author.

Figure 1.1 Representations of the process of policy – research feedback loops

of research being adopted by policy makers in development: the *political context* defined by the degree of openness and political freedoms as well as local history and vested interests; *evidence* referring to the quality, usefulness and problem-solving properties of research as well as how it is packaged and presented; *links* between the research and policy communities in affecting policy formulation; and, finally, *external influences*, such as donor strategies and international politics and processes.

This analytical framework points to a dynamic relationship between researchers and policy makers. In a well-functioning relationship, the activities of the two sets of actors mutually reinforce the activities of the other in a positive feedback loop as depicted in the left-hand side of Figure 1.1. An innovation system where such a dynamic interaction takes place has – in our terminology – been successful in Closing the Loop.

In cases where, for reasons discussed above, the feedback mechanisms do not function properly, the process of research-to-policy and policy-to-research is dysfunctional and cannot provide either set of actors with the needed information (the loop is not closed).

So how can a virtuous cycle of mutual reinforcement between researchers and policy makers be achieved? It would seem that most countries in the world are situated somewhere between the ideal state and the system breakdown represented by the two flowcharts in Figure 1.1 and struggles with improving the dynamics. However, if it can be difficult for developed countries, it can be harder still in developing countries. Carden (2009) lists a host of problems why policy making in developing countries is harder than in developed countries. Chief among them are:

- policy makers have less autonomy
- staff turnover weakens both research and policy influence
- lack of intermediary institutions
- implementation challenges are greater
- 'personal' relationships can lead to misgovernment
- policy makers lack confidence in their own researchers
- researchers in development often lack hard data
- demand for research can be missing.

Addressing these problems is a formidable task for any country. Luckily, all barriers are not present at the same time and every country has its own path to follow and ways to cope with these issues. Two key requisites for moving forward are that there is a genuine interest from both sides of the two sets of actors and that proper mechanisms for dialogue and communication between them exist. Figure 1.2 describes the basic tools available for researchers and organizations trying to impact policy formulation. The methods range from purely cooperative measures to outright confrontation through direct action. In the narrative of this book we shall be exploring primarily the upper left-hand side of the figure; a cooperative relation providing evidence-based and science-based advice through policy

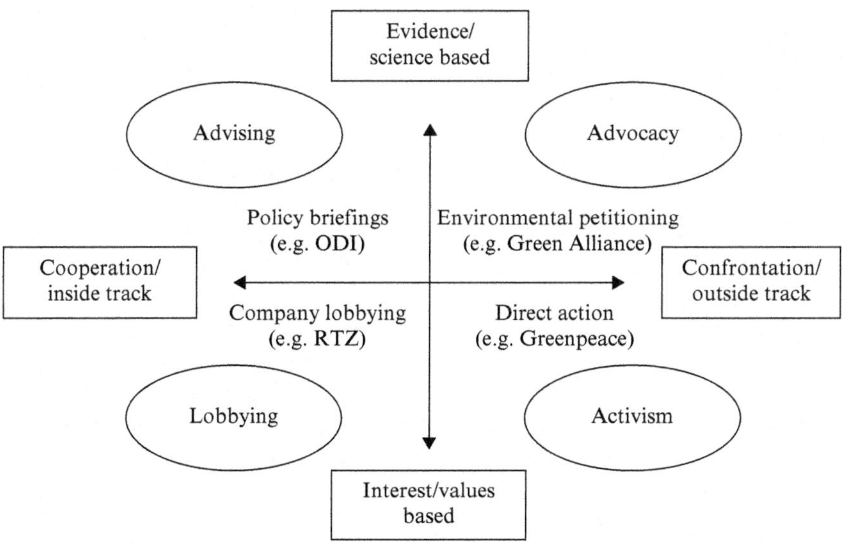

Source: Start and Hovland (2004).

Figure 1.2 Tools and organizations on the cooperation/evidence axes

briefings and other research outputs for decision makers. Thus, we are concerned with the research community's impact on the policy-making process as well as the dynamic interaction between research and policy.

The other side of the loop, how policy can influence research, is somewhat more straightforward – but no less controversial – given the sway policy makers typically hold over the funding of research. By providing earmarked grants for strategic or needs-driven research, funding agencies can influence the research direction as well as size. A global trend in research funding is the decrease of core funding for universities and research institutes in favour of external, project-based funding from research councils and foundations. However, as we discussed above, the internal academic incentive system may not always be in tune with the demands of the granting agencies.

1.4 STRUCTURE AND CONTENT OF THE BOOK

In Chapter 2, the socio-economic context for the study is presented. The chapter discusses the recent developments in the case countries of using the Millennium Developmental Goals (MDGs) as a yardstick for how well attuned to the development objectives are the national systems of innovation. It provides the background to the country case studies presented in the following chapters by examining the progress made in meeting socio-economic goals such as eradicating extreme poverty and diseases like HIV/AIDS and malaria, lowering infant and maternal mortality rates, increasing literacy and proving universal primary education, and promoting gender equality. All four countries have made considerable progress in meeting the MDGs although Bolivia and Vietnam have done relatively better than Mozambique and Tanzania. On the other hand, it is shown that Mozambique and Tanzania have been more successful than many sub-Saharan African countries.

Chapter 3 traces the origins of the national innovation systems (NIS) approach and discusses how the concept has been incorporated and applied in the context of developing countries. The emergence of NIS in developing countries is discussed and exemplified by tracing the history of STI policies and institutional infrastructure in the four case countries (Bolivia, Vietnam, Tanzania and Mozambique). The evolution of the NIS in the case countries indicates that although they are organically different and path dependent, there are similar features and common lessons to be drawn from their respective experiences.

Bolivia was early in recognizing the role of science and technology (S&T) in development, as was indeed the whole Latin American region. The Latin

American School of Thought in Science and Technology emerged in the 1950s when many Latin American countries initiated the process of incorporating S&T into national development plans. Bolivia was one of the first countries in the region to establish a policy-making body by the founding of the National Academy of Sciences as an autonomous public organization in charge of the formulation of science policy. Moreover, Bolivia was the first country in the region to formally make the system's approach operational. The emergence and growth of the institutional set-up was not problem-free. Turf disputes resulting from overlapping or unclear areas of responsibility between leading institutions in STI policy formulation and knowledge production have periodically hampered the efficient functioning of the system. To some degree this still persist, with science policy agendas and research agendas being set by different agents of the Bolivian innovation system.

For Vietnam, the emergence and development of an innovation system took place over a period of several decades, with the concept of innovation slowly but gradually becoming integrated into mainstream thinking in policy-making circles. The work on building up S&T capacities in the 1960s provided the basis for the gradual evolution into an innovation system approach in the 1990s. In this process, collaboration with foreign partners has proved instrumental for the policy-making process in the country by providing new analytical tools and frameworks. In particular, the major shift from a preoccupation with S&T to innovation aspects, as clearly seen today, owes much to this collaboration. Vietnam also differs from the other case countries in that in the 1980s the country went through a complete overhaul of the system of managing the S&T sector. The Soviet style of organizing S&T activities was gradually changed and replaced by new institutions, laws and policies allowing the country to adopt more of a mainstream thinking in NIS policy making and more active integration in the international market system.

Tanzania was also early in recognizing the importance of developing policies for guiding S&T efforts in the country and actively participated in international and regional attempts to promote the understanding and usefulness of S&T policies for development. Despite the early insights into the role of S&T, the historical record has been rather disappointing both in terms of policy consistency and the application of the results of S&T where the achievements of the sector have not been translated into tangible products, processes and services for development purposes. A large number of factors still impede the efficient creation and utilization of new knowledge, such as a weak multi-disciplinary interaction and collaboration between R&D organizations, weak social and economic research, inadequate mechanisms for including new and emerging research areas into the national research agenda and low participation of the private sector in research activities. These limitations are further exacerbated by inadequate human resources and an over-dependence on foreign funding for research. Moreover, as in the case of Bolivia, Tanzania has experienced its share of turf wars between competing STI organizations with overlapping areas of responsibility. The latest National Research and Development Policy is designed to address these shortcomings by strengthening the research capacity, establishing an appropriate coordination and management system of research activities and setting clear and realistic priorities for research on short-, medium- and long-term bases.

Although Mozambique, as indeed most African countries, early on acknowledged the importance of S&T for development, it was not until after the end of the devastating civil war in 1992 that work on building up the NIS could start. Thus, the emergence of the innovation system approach in Mozambique came some time later than in many developing countries. It was triggered by extensive discussions at the national and regional levels and the contribution of several national researchers' work. Being late can work as an advantage in the sense that it is possible to leap-frog to a well-defined policy leading the development of the innovation system by taking advantage of the experiences of other countries and avoiding costly mistakes. Mozambique has introduced definitions and concepts from more developed countries and added some local ingredients, such as the case of the local communities as strong actors. In doing so, and unlike other countries in Africa, Mozambique has a very solid potential base for its build-up.

However, despite the government's efforts in developing various initiatives to implement S&T policy, establish a legal framework and mobilize financial resources, the Mozambique innovation system is still new and much effort needs to be made in order for it to operate successfully. Chief among these are issues related to linkages between the different actors of the system as well as the financial and legal autonomy of the research institutes.

The overviews of the emergence of NIS in the four countries indicate that the process has been both long and contentious. Bolivia, Tanzania and Mozambique all experienced conflicts to differing degrees with the establishment of a Directorate or Ministry of Science and Technology taking over responsibility for the area of STI. Already established institutions fought the perceived usurper; in Bolivia at the end of the 1970s, in Tanzania in the 1980s and in Mozambique in the 2000s. In Vietnam, the Ministry of Science and Technology does not seem to have met such resistance given the 'silo-structure' where each ministry has its own funds for R&D, but the creation of the National Council for Science and

Technology Policy in 2003 seems to indicate the need to sort out areas of responsibility and jurisdiction in STI in Vietnam.

After the historical overview of the emergence of the NIS in developing countries discussed in Chapter 3, the following four chapters go deeper into the present state of the innovation system in the case countries. First, each of these chapters discusses the main properties of the NIS and analyses the structural and organizational weaknesses of the innovation system as well as the main challenges for the system. Second, the chapters focus on the sectoral innovation systems for health and agriculture and analyse the results of a survey on experiences of researchers in aligning project work with overall development goals and in influencing policy.

Chapter 4 discusses the case of Bolivia. It is argued that 'Closing the Loop' in Bolivia is indeed possible as there exists considerable human potential for obtaining research outputs for social and economic innovations. There is, however, a set of conditions that must be met in order to complete the process, in particular the need to adopt a more flexible evidence-based policy, allowing all actors to support the co-evolution of all the elements that make up the NIS.

STI policies are only effective when government can formulate and implement them. In Bolivia it is clear that the main reason for the difficulty in building up an adequate capacity has been insufficient government capabilities and effectiveness. Thus, a stronger compromise of the State is needed with the NIS in the way of increased resources and infrastructure.

Several indicators show that the NIS is at present weak in Bolivia. The causes for this situation are well known, and discussed in depth in Chapter 4: lack of financial inputs, limited efforts of the business sector, weak human resource base and so on. However, talent exists and significant research outputs from public and private research centres are diffused through scientific publications and technical reports as well as in relation to specific economic and social developments. In recent years, private sector associations have become more active in promoting technology and innovation within their networks and together with public and private universities.

The systems of innovation in the sectors analysed operate at varying degrees of efficiency. As discussed in the chapter, the traditional agriculture production model has indeed produced innovations, but is criticized by policy makers as not having valued the local knowledge and wisdom of indigenous people, nor the wealth of the existing biodiversity. This new approach in Bolivia distances itself from an innovation strategy oriented towards competitiveness, value chains and productive specialization and instead tries to construct sustainable management of natural resources and the promotion of biodiversity based on an endogenous development model.

The health system has always been considered a priority in Bolivia and consequently much effort has been placed on research, however falling short of innovations. The health innovation system is today composed of a large number of public and private organizations and a research agenda has been defined.

Research into renewable energies has been conducted for several years and there are already research outputs that have 'Closed the Loop' and can serve as benchmarks. Research and innovation in the informal sector have not been analysed in detail, although characterizations have been made in the studies on small and medium-sized enterprises (SMEs) and training and innovation requirements have been identified. This latter sector is key to the Bolivian economy.

In the sectors analysed it is found that 'Closing the Loop' has been possible when dealing on a product-by-product basis. Considering existing experiences, it is clear that selective interventions must be sought that focus on strategic areas that can help accumulate the know-how and capabilities for further refinement. It is pointed out that although alternative approaches are important they cannot be the sole approach, as suggested by the proponents of an endogenous development model. An overemphasis on such a model disregards conventional approaches that have already been shown to work.

In Vietnam, the alignment of development plans with the activities of knowledge producing organization is quite strong. As deliberated in Chapter 5, R&D institutes, universities and even some companies have intentionally oriented their missions to address societal needs in development. It appears that at least some of them have sufficient capabilities to deliver relevant scientific research results as well as evidence-based support to policy making. Working closely with farmers is one of the most preferred modes of operation in order to ensure that needs in agriculture or health are properly addressed. However, there are severe policy and institutional problems that hinder the full potential of this process from to be reached.

The government has adopted a policy of creating a selective number of strong public universities in an attempt to create 'centres of excellence' for university education and research. However, the quality of research in Vietnam is still comparably low with a number of weaknesses and shortcomings. Wars and isolation from the international scientific communities have hampered the gradual build-up of physical infrastructure and human resources. Many universities are far from being considered R&D centres of excellence and lack an autonomous status to pursue their own strategies for quality improvements. Their staff, especially in public universities, face constraints in terms of salary ceilings, human resource management regulations and financial incentives set by the government. Despite the introduction of research as a complementing mission at universities, there is still a well-documented lack of research and weak linkages between research and teaching. Moreover, the present incentive scheme does not promote collaboration within the university, nor does it provide mechanisms to encourage interaction with other institutions and firms. Cooperation is usually short term, and relies mainly on personal and informal relationships. Consequently, the relevance of much of the R&D carried out at universities and research institutes may be called into question.

In the business sector, the technology level of the enterprises is in general outdated and the capacities for adopting advanced technological equipment rather limited. This applies particularly for SMEs, but large state-owned enterprises also tend to pay insufficient attention to the search for technical information, improving S&T knowledge and investing in technology innovation. Enterprises do not express long-term visions, having little interest in technology innovation for sustainable development with a consequent low rate of investment in new technology.

In spite of such shortcomings, S&T has made remarkable contributions to economic growth and social development in Vietnam. Specific cases presented in the chapter reveal that many organizations were indeed able to deliver what they were expected to do in terms of producing relevant research results and training for development needs. In some instances, the achievements are quite extraordinary like the self-sufficiency of vaccines for national vaccination programmes or new species and crops in rice and maze production.

For such achievements to become the norm rather than the exception, and for the relevance of the R&D sector in Vietnam to improve, the responses to the survey of researchers offer some clues. One important factor to emerge is the fact that development goals are sometimes perceived to be formulated in too general and broad terms, often resulting in a R&D agenda based on the specificity of the involved organizations, particularly the ones who fund the research, rather than in compliance with overall developmental goals. The survey also reveals that a systematic structure to support a dynamic interaction between researchers and policy makers is sorely lacking despite some recent initiatives.

In Tanzania a poor record of knowledge-to-policy can be observed. In Chapter 6, the overall assessment points to the fact that the NIS in Tanzania is in grave disorder despite examples of isolated best practices and successful cases. It would appear that the way science-based research is designed, implemented and applied in the country does not easily lend itself to 'Closing the Loop'.

An important feature of the research environment in Tanzania is the dominating position of the international donor community in research funding, with the majority of the research in the country being funded by external sources. Obviously this impacts the way research is formulated and conducted and there is a general perception in the country that donor funding is bad because it does not focus squarely on national development goals. The survey confirms that this is a prevalent notion, although it also notes that there might be large differences between scientific fields – in agriculture the majority of the researchers feel that donor funding comes with objectives not necessarily tied to national development plans, whereas in the health sector only one in four researchers expressed such sentiments. In this case, the universality of global health issues appears to better align donor funding in the health sector with development plans compared with the specificity and local requirements of other research areas.

The survey identifies additional challenges facing the research community, including a paucity of funds from government and donors, limited opportunities for collaboration with other researchers in the region or sub-region, inadequate protection of intellectual property rights and a persistent gap in collaboration between researchers, policy makers and end users of research results.

Remedying this situation requires action on several levels. To align knowledge generation to the social and economic needs of the country, the respondents to the survey expressed a desire for a clear national research agenda developed sector-wise, coupled with an adequate budget. To ensure that the generated knowledge is relevant in meeting national goals, research projects should be negotiated with the users during the design stage. Most importantly, mechanisms for presenting research results in non-confrontational ways need to be developed.

There are some positive signs of change. The government has recently decided to increase funds for research and the Commission for Science and Technology, which is responsible for national research priority setting, is currently reviewing its national priorities for research in different sectors. Responses to the survey indicate that while these developments are indeed good news, they are not a panacea. If Tanzania's stated ambition to become a middle-income country by 2025 through the application of science, technology and innovation is feasible or even credible, the restrictions facing the research community must be addressed and overcome.

Of the four country cases discussed in this book, Mozambique is the poorest in terms of GDP per capita and also has the least developed NIS. Chapter 7 discusses how the main coordinating body for S&T-related activities in Mozambique, the newly restructured and re-named Ministry of Science and Technology, Higher, Technical and Professional Education and its predecessor Ministry of Science and Technology (MCT), have struggled to create a supportive infrastructure and collaborative environment for the actors of the NIS. Despite some encouraging progress, the overall results have been meagre. One reason is that of the unresolved issues of areas of responsibility of the Ministry. It is argued that turf wars and conflicts of interest between MCT and other ministries and public institutions have seriously hampered the implementation of the ten-year master plan for science and technology (MOSTIS) adopted by the government in 2008. The absence of coordinated efforts, coupled with a lack of metrics for measuring outcomes and outputs and indicators for attainment of targets, constitute major obstacles for the implementation of MOSTIS and an efficient S&T policy.

A second problem in the implementation of MOSTIS is the lack of continuous updating and scrutiny of activities under the plan. Some parts of the strategic programmes have become unsuitable under present conditions while others have not even been implemented since 2008. Furthermore, the plan for enhancing the NIS rests on the assumption that the participating institutions are bestowed with appropriate financial and human resources. Evidence shows that most institutions do not have those prerequisites and lack a clear understanding of what they are tasked with in the field of STI.

To compound the situation there is a general lack of qualified human resources in the NIS. Systematic inclusion of science, technology and innovation issues in developmental plans is relatively new in Mozambique and many planners, policy makers and actors of the NIS demonstrate difficulties in understanding and implementing the sometimes complex programmes envisaged in MOSTIS. This is also evidenced by the lack of collaboration among different institutions of the NIS and by the weak institutional linkages between them.

In spite of systemic and management problems, relevant research is carried out by universities and specialized research institutes. However, as the survey of researchers in Mozambique indicates, most of the technology developed in the agrarian sector is not used or delivered properly to the target group, the farmers, due to systemic failures. A complicating factor is the heavy reliance on international donors for the funding of research projects with over 70 per cent of gross expenditure on R&D in Mozambique coming from international sources. Many of the researchers feel that the interests and visions of external funders are not always compatible with national development plans. In order to qualify for competitive donor grants, it is important to obey the criteria, rules and most important the research interests and area set by the funder, which may not always coincide with the national programmes.

The final chapter provides a synthesis of the empirical results discussed in Chapters 2 to 7. The surveys of the perception of the researchers on the functioning of the NIS in the four countries are analysed and compared. The respondents' perceptions of quality, relevance and impact of research point to the fact that the research communities in these countries are both willing and able to contribute to generating relevant research results and informing policy. However, the results of the surveys also point to a systemic failure in the ability of the NIS to make this happen.

NOTE

1. The concept of 'Closing the Loop' originates from the International Development Research Centre (IDRC), which defines it as 'an approach to programming and projects that seeks to ensure the awareness, understanding, and ownership of research outputs by decision-makers at all levels. Its goals are to increase the relevance and utilization of research outputs, thereby enhancing the influence of the researchers, institutions, and work we support' (IDRC, 2002, p. 1).

REFERENCES

Carden, F. (2004), 'Issues in assessing the policy influence of research', *International Social Science Journal*, **56**(179), 135–51.

Carden, F. (2009), *Knowledge to Policy. Making the Most of Development Research*, New Delhi: Sage Publications/IDRC.

Court, J. and J. Young (2006), 'Bridging research and policy in international development: an analytical and practical framework', *Development in Practice*, **16**(1), 85–90.

Dosi, G., P. Llerna and M.S. Labini (2005), 'Science-technology industry links: "the European paradox" – some notes on the dynamics of S&T research in Europe', LEM Working Papers, Laboratory of Economics and Management, Pisa.

Ejermo, O. and A. Kander (2006), 'The Swedish paradox', CIRCLE Working Paper Series, 2006/01, Lund University.

GFHR (Global Forum for Health Research) (2004), *The 10/90 Report on Health Research 2003–2004*, Geneva: Global Forum Publications.

Hy, R.J., M. Venhaus and R.G. Sims (1995), 'Academics in service to the legislature: legislative utilization of college and university faculty and staff', *Public Administration Review*, **55**(5), 468–74.

IDRC (2002), *Closing the Loop – Communication for Change at IDRC*, available at https://idl-bnc.idrc.ca/dspace/bitstream/10625/21181/1/117024.pdf (accessed 3 May 2015).

Jacobsson, S. and A. Granberg (2006), 'Myth and reality – a scrutiny of dominant beliefs in the Swedish science policy debate', *Science and Public Policy*, June, 321–40.

Jefferys, M., K. Troy, N. Slawik and E. Lightfoot (2007), 'Issues in bridging the divide between policymakers and researchers', University of Minnesota, available at http://cascw.umn.edu/wp-content/uploads/2014/07/BridgingDivide-PolicyResearch.pdf (accessed 3 May 2015).

Lundvall, B.A., K.J. Joseph, C. Chaminade and J. Vang (2009), 'Innovation systems and developing countries – an introduction', in B.A. Lundvall, K.J. Joseph, C. Chaminade and J. Vang (eds), *Innovation Systems and Developing Countries – Building Domestic Capabilities in a Global Setting*, Cheltenham, UK and Northampton, MA, USA: Edward Elgar.

Mazzucato, M. (2013), *The Entrepreneurial State. Debunking Public vs. Private Sector Myths*, London: Anthem Press.

Moravcsik, M. (1966), 'Some practical suggestions for the improvement of science in developing countries', *Minerva*, **4**, 381–90.

Piketty, T. (2014), *Capital in the Twenty-first Century*, Cambridge, MA: Harvard University Press.

SPP (2009), *Science and Public Policy*, **36**(2), March, Special issue on the Third Mission of Universities.

Start, D. and I. Hovland (2004), *Tools for Policy Impact: A Handbook for Researchers*, London: Overseas Development Institute.

Surr, M., A. Barnett, A. Duncan and M. Spreight (2002), 'Research for poverty reduction', DFID Research Policy Paper, London.

2. The socio-economic context and the Millennium Development Goals

Claes Brundenius

2.1 POPULATION CHARACTERISTICS

The countries in the survey, Bolivia in South America, Mozambique and Tanzania in East Africa and Vietnam in Southeast Asia, have many similarities but they are also different in many aspects. They vary in size both with respect to area extension and population. All the countries are large by surface area, with Bolivia being by far the largest, followed by Tanzania, Mozambique and Vietnam (Table 2.1). Surprisingly perhaps, we find that Bolivia is by area-size over three times larger than Vietnam. With respect to population the situation is quite different, with Vietnam having by far the largest population (88.8 million inhabitants), followed by Tanzania

Table 2.1 Size of the countries and population characteristics, 2013

	Bolivia	Mozambique	Tanzania	Vietnam
Surface (000 sq km)	1098.6	799.4	947.3	331.0
Population (million)	10.5	25.2	44.9	88.8
Population density	9.6	31.5	47.4	268.3
Population growth	1.7%	2.8%	2.7%	1.0%
Population below 15	35%	45%	45%	23%
Population 15–64	60%	51%	52%	71%
Population 65+	5%	5%	3%	7%
Age dependency[a]	66%	95%	93%	41%
Median age	22.2	17.5	17.3	27.4

Note: a Age dependency ratio is the ratio of dependents – people younger than 15 or older than 64 – to the working age population (as a percentage of the working population).

Source: World Bank (2015) and GPO (2014).

(44.9), Mozambique (25.2) and last Bolivia with 10.5 million. Bolivia has thus the lowest population density of the countries with only 10 people per square kilometre, with Vietnam at the other extreme with 268 people per square kilometre. Population growth is high in Mozambique and Tanzania – 2.5 and 2.8 per cent, respectively – while modest in Vietnam (1.0 per cent), and also not so high in Bolivia (1.7 per cent). The population estimates for Mozambique and Tanzania take into account death rates as a result of AIDS that has dramatically affected sub-Saharan Africa, and Southern and Eastern Africa in particular.

Mozambique and Tanzania have a very young population (with a median age of around 17 years) although this is compensated somewhat by a rather low share of older people in the population (more than 65 years of age). The age dependency ratio is quite high in the two African countries.

2.2 LEVELS OF DEVELOPMENT AND ECONOMIC GROWTH PATTERNS

Although all four countries are relatively poor, they are yet at different levels of development (Table 2.2). While Bolivia has been a lower middle income country for a long time, it is only recently that Vietnam has also entered this income group (according to the United Nations Development Programme (UNDP) definitions). Mozambique and Tanzania both belong to the low income group (or the least developed countries). Bolivia and Vietnam have a per capita income that is about three times higher than Tanzania and four times higher than Mozambique. Bolivia thus has the highest per capita income of the four countries (4880 USD (PPP, purchasing power parity) in 2013) in spite of being the poorest country in South America (Haiti being the poorest country in Latin America and the Caribbean).

The patterns of growth of the countries are quite different. Figure 2.1 gives a good illustration. It depicts the long-term growth trend of per capita income in 1970–2014. The long-term trend over these 44 years is very slow growth in Bolivia, only 1.1 per cent per annum, but the trend is also slow growth in Mozambique and Tanzania (2.4 per cent and 2.1 per cent per annum, respectively). In contrast, Vietnam's growth record in the same period is quite impressive, with an average growth of 4.6 per cent.

All four countries experienced slow growth in the 1970s, including recession during parts of the 1980s. This trend continues also in the 1990s, with negative growth in both Bolivia and Tanzania. In Vietnam growth begins to pick up after the economic reforms started at the end of the 1980s, but

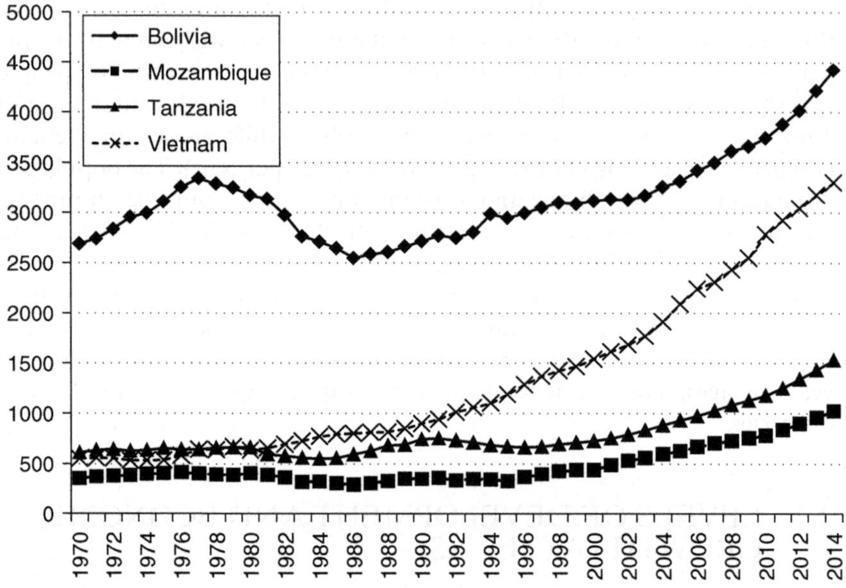

Source: PWT (2015), extrapolations: ECLAC (2015), AFDB (2015) and GPO (2014).

Figure 2.1 Long-term trend: per capita economic growth in Bolivia, Mozambique, Tanzania and Vietnam (2005 USD, PPP), 1970–2014

especially after 1990 when serious structural market reforms are carried out. The 1990s is the decisive decade for Vietnam initiating a period with high and sustained growth, contrasting with the stagnation and recession in Bolivia, Mozambique and Tanzania in the same decade.

The picture changes radically in the first decade of the new millennium. Growth in sub-Saharan Africa picks up and the average per capita growth in Mozambique and Tanzania during the last 10–15 years has been among the highest in the world. But let's not forget that the increase starts from very low levels of income. It seems that Bolivia has at last come out of a long secular stagnation that has lasted since the 1950s when Bolivian gross domestic product (GDP) per capita (3130 USD, constant PPP) was not far from that of Chile (3250 USD, constant PPP) (PWT, 2015).

The industrial sector is more developed in Bolivia and Vietnam in relation to the African countries, with a share of 38 per cent d GDP. The service sector comprises about half the economy in all four countries (Table 2.2).

Table 2.2 Economic performance and structural differences, 2012/2013

	Bolivia	Mozambique	Tanzania	Vietnam
GNI/capita (USD)	2220	510	570	1550
GDP/capita (USD, PPP)	4880	1000	1560	3620
Average annual per capita growth rates:				
1970–80	1.7%	1.4%	0.8%	1.4%
1980–90	−1.6%	−1.5%	1.1%	3.6%
1990–2000	1.4%	2.3%	−0.3%	5.5%
2000–10	1.8%	6.0%	5.0%	6.1%
2010–14	4.3%	6.3%	6.8%	4.4%
1970–2014	1.1%	2.4%	2.1%	4.2%
Economic sector shares:				
AGR VA % GDP	13.3%	29.0%	28.4%	18.4%
IND VA % GDP	38.1%	20.8%	24.4%	38.3%
SERV VA % GDP	48.6%	50.2%	47.3%	43.3%

Note: GNI is gross national income.

Source: UN (2015a, 2015b).

2.3 HEALTH SITUATION

Health indicators are closely related to the development levels of the countries (Table 2.3), with Vietnam at the top. Life expectancy is thus highest in Vietnam, almost at the level of a developed country (76 years), considerably lower in Bolivia (67 years) and in sharp contrast with Tanzania (61 years) and Mozambique (50 years). The same pattern is found with respect to infant mortality rates and mortality under five years of age. Vietnam also has quite lower maternal mortality than the rest of the countries. One reason could be the relatively higher density of physicians in Vietnam: 12 physicians per 10,000 inhabitants, compared to around 5 physicians in the other three countries.

Public hospitals are important in all four countries, but especially in Bolivia where public health expenditures make up around 78 per cent of the total, while this share is only around 40 per cent in the other three countries. This is also reflected in the government budgets, where Bolivia

Table 2.3 Health indicators, 2012/2013

	Bolivia	Mozambique	Tanzania	Vietnam
Life expectancy	67	50	61	76
Infant mortality per 1000 live births	31	62	36	19
Physicians per 10,000 inhabitants	5	4	5	12
Health expenditures (% government expenditure)	16.4%	7.5%	11.2%	6.0%
Total health expenditures (% of GDP)	5.8%	6.4%	9.6%	6.6%
Public health expenditures (as % of total health expenditures)	78.4%	46.4%	36.3%	41.9%

Source: World Bank, World Health Organization (WHO) and Hanoi Government Printing Office (GPO).

devotes about 16 per cent of the budget to health expenditures, compared to 11 per cent in Tanzania and 6–7.5 per cent in Mozambique and Vietnam.

2.4 PROGRESS IN PUBLIC EDUCATION

All four countries in this survey devote considerable resources to education (Table 2.4). Tanzania spends almost 28 per cent of its budget on education, implying that public expenditures on education constitute 6 per cent of GDP. Mozambique spends relatively less on education than the other there countries. This is reflected in considerably lower literacy rates in Mozambique – especially with respect to the elderly population.

All countries keep practically all their children, in corresponding ages, in primary schools. With respect to secondary schools the picture is quite different. Bolivia and Vietnam have higher enrolment rates (around 75 per cent) than Mozambique and Tanzania (26 per cent and 35 per cent, respectively). The contrast between Bolivia/Vietnam and Mozambique/Tanzania is even higher when it comes to the university level.

Table 2.4 Education indicators, 2012/2013

	Bolivia	Mozambique	Tanzania	Vietnam
Public expenditure EDU (%) of total government expenditure	19.5%	18.1%	27.5%	20.9%
Public expenditure EDU (%) of GDP	6.9%	4.1%	6.2%	6.3%
Literacy rates youth (15–24)	99.0%	67.2%	74.6%	97.1%
Literacy rates adult (15+)	94.5%	50.6%	67.8%	93.5%
Literacy rates elderly (65+)	72.3%	21.2%	52.4%	76.9%
Enrolment ratios at primary level (%)	94%	105%	93%	105%
Enrolment ratios at secondary level (%)	77%	26%	35%	76%
Enrolment ratios at tertiary level (%)	34%	5%	4%	25%

Source: UIS (2015).

2.5 HUMAN DEVELOPMENT INDEX AND INCOME GAPS

The UNDP annually publishes its *Human Development Report* (HDR), which with its Human Development Index (HDI) gives rough estimates of the level of living for practically all countries, levels that are not necessarily equivalent to their GDP (or GNI) per capita levels. The data for the survey countries, as reported in the latest HDR (UNDP, 2015), are shown in Table 2.5. Bolivia ranks highest (113) with Vietnam close behind (121). Tanzania and Mozambique rank much further behind: 159 and 178, respectively (of 187 countries). Bolivia and Vietnam are in this ranking classified as Medium Human Development countries, while both Mozambique and Tanzania are classified as Low Human Development countries.

There had been a lot of doubt about the usefulness of the HDI since the GDP per capita did not consider income inequalities. In 2013 the HDR therefore included an index adjusted for income inequality (IHDH). This adjusted index does change the picture to some extent as a result of the Gini coefficients. With the IHDI Bolivia loses 10 ranks in comparison with HDI, while Vietnam gains 15, Tanzania 8 and Mozambique 2.

Table 2.5 Human Development Index and poverty reduction, 2014

	Bolivia	Mozambique	Tanzania	Vietnam
HDI (rank)	0.667 (113)	0.393 (178)	0.488 (159)	0.638 (121)
HDI inequality adjusted (difference from HDI rank)	0.470 (−10)	0.277 (+2)	0.356 (+8)	0.543 (+15)
GINI coefficient	0.466	0.457	0.376	0.356
Share in income of 10% richest	33.6%	36.7%	29.6%	27.8%
Share in income of 10% poorest	0.8%	1.9%	2.8%	2.9%
People below 2 USD (PPP) per day	13%	82%	73%	12.5%
People below 1.25 USD (PPP) per day	8%	61%	44%	2.4%

Source: World Bank (2015) and UN (2015b).

Bolivia has the most unequal income distribution (one of the worst in Latin America) with a Gini of 0.47 followed by Mozambique, Tanzania and Vietnam. This inequality is also reflected in the income shares accrued to the highest deciles and lowest deciles. The income distribution is slightly better in Tanzania and Vietnam in comparison with Bolivia and Mozambique.

2.6 MILLENNIUM DEVELOPMENT GOALS AND ACHIEVEMENTS

In 1990 the General Assembly of the United Nations (UN) set up the so-called Millennium Development Goals (MDGs), ambitious global goals to be reached by 2015 in a series of areas, such as eradicating extreme poverty and diseases (like HIV/AIDS, malaria), lowering infant and maternal mortality rates, increasing literacy and providing universal primary education, and promoting gender equality and empowering women.

In July 2015 the UN released *The Millennium Development Goals Report 2015* (UN, 2015a). The report presents some preliminary estimates on how developing countries have fared in achieving the 2015 targets. In the following we present the official UN global data including the data for the four survey countries, indicating to what extent the goals/targets have been achieved. The 2015 data are official UN estimates. We are using the MDGs here as proxies for how well the national innovation system is aligned with the development goals of the case countries.

2.6.1 Goal 1: Reduce Extreme Poverty

The first and perhaps most important goal of the MDGs is to eradicate extreme hunger and poverty. Its Target 1.A has been to cut by half the proportion of the population living on less than 1.25 USD (PPP) a day. This is the extreme poverty line. Another measure of extreme poverty is the percentage of the population that is undernourished. The achievements are shown in Tables 2.6 and 2.7.

The overall achievement for developing countries is quite astounding. In developing countries, including China, the number of extremely poor diminished from almost half to 14 per cent, between 1990 and 2015. This means that the target has been achieved by a large margin. However, it should be stressed that to a great extent this success is due to China where millions of people have been moved out of poverty during the last 25 years. As *The Milllennium Development Goals Report 2015* states, 'China alone

Table 2.6 Proportion of people living on less than 1.25 USD (PPP) a day (%)

	1990	2015	Target achievement
Bolivia	5	8	Negative
Mozambique	81	61	20% from target
Tanzania	72	44	8% from target
Vietnam	64	2.4	Target achieved
Memo items:			
sub-Saharan Africa	57	41	12% from target
Southeast Asia	46	7	Target achieved
China	61	4	Target achieved
Latin America and the Caribbean	13	4	Target achieved
Developing countries (incl. China)	47	14	Target achieved
Developing countries (excl. China)	41	18	Target achieved

Source: UN (2015a, 2015b).

Table 2.7 Proportion of undernourished people (%)

	1990	2015	Target achievement
Bolivia	38	16	Target achieved
Mozambique	56	25	Target achieved
Tanzania	24	32	Negative
Vietnam	46	11	Target achieved
Memo items:			
sub-Saharan Africa	33	23	7% from target
Southeast Asia	31	10	Target achieved
East Asia	23	10	Target achieved
Latin America	14	5	Target achieved
Developing countries	23	13	Target achieved

Source: UN (2015a, 2015b).

accounts for almost 2/3 of the total reduction in the number of undernourished people in developing countries' (UN, 2015a).

Three of the four survey countries have also decreased extreme poverty levels, most dramatically so in Vietnam. Extreme poverty has also decreased in both Mozambique and Tanzania, although not by half. In sub-Saharan Africa the number of extremely poor actually increased in both absolute numbers (headcounts) and as a percentage of the total in the 1980s, from 214 to 319 million, and from 54 per cent to 58 per cent, respectively (Paus, 2013, table 1.1). The trend is fortunately reversed after 2005. The percentage living in extreme poverty in sub-Saharan Africa decreased from 51 per cent in 2005 to 41 per cent in 2015. However, in spite of this relative success, the absolute number of extremely poor has decreased only slightly from 391 to 389 million.

In Bolivia the percentage of extremely poor has, on the other hand, increased, albeit from a low level, 5 per cent in 1990 to 8 per cent in 2015. This means of course that absolute numbers (headcounts) have increased as well. The most dramatic reduction of extreme poverty has taken place in Vietnam. Between 1990 and 2015, the percentage of extremely poor reduced from 61 per cent to 2.4 per cent, which means in total numbers from 42.2 million people to 2.3 million (GPO, 2014; UN, 2015b).

Another sub-target under this first goal has been Target1C: to 'reduce by half, between 1990 and 2015, the proportion of people of undernourished people'. The results are shown in Table 2.7. But then who is undernourished? The MDG is using the Food and Agriculture

Organization (FAO) definition, which is the proportion of the population below the minimum level of dietary energy consumption, as a percentage of the total population (for details, see FAO, 2015).

As seen in Table 2.7, the targets have been achieved practically in all developing regions. Quite a success story! Sub-Saharan Africa is close to the target. The target has been achieved in three of the survey countries. The only exception is Tanzania, where the proportion of undernourished has actually increased. This appears to contradict the findings in Table 2.6, showing that the number of Tanzanians living in extreme poverty has fallen from 72 per cent to 44 per cent. On the other hand, the *Tanzania Human Development Report 2014* (UNDP, 2014) seems to confirm some of the negative trends.

2.6.2 Goal 2: Achieve Universal Primary Education

The most important target here is to 'ensure that, by 2015, children everywhere, boys and girls alike, will be able to complete a full course of primary schooling'. This of course means that that by 2015 all (100 per cent) children should complete primary schooling, quite an ambitious target. In fact it is only in East Asia that the target has been reached. Among the survey countries it is only Vietnam that achieved the target. In Bolivia the result is surprisingly negative, with less enrolment ratios in 2015 compared to 1990. On the other hand, both Mozambique and Tanzania have been quite successful with increasing enrolment ratios (Table 2.8).

Table 2.8 Net enrolment ratios in primary education (%)

	1990	2015	Target achievement
Bolivia	95	82	Negative
Mozambique	44	88	Close to target
Tanzania	51	85	Close to target
Vietnam	98	98	Target achieved
Memo items:			
sub-Saharan Africa	52	80	Positive
Southeast Asia	93	94	Very close to target
East Asia	97	97	Target achieved
Latin America and the Caribbean	87	94	Very close to target
Developing countries	80	91	Close to target

Source: UN (2015a, 2015b).

Table 2.9 Gender disparity in primary, secondary and tertiary education (females as % of males)

	1990			2015			Target achievement		
	Prim.	Sec.	Tert.	Prim.	Sec.	Tert.	Prim.	Sec.	Tert.
Bolivia	0.93	0.93	0.84	0.97	1.00	0.84	Yes	Yes	No
Mozambique	0.76	0.57	0.32	0.91	0.91	0.69	Close	Close	No
Tanzania	0.99	0.75	0.19	1.04	0.92	0.54	Yes	Close	No
Vietnam	0.99	0.95	0.66	0.98	0.95	0.90	Yes	Close	Close
Memo items:									
sub-Saharan Africa	0.83	0.76	0.54	0.93	0.88	0.62	Close	No	No
Southeast Asia	0.95	0.91	0.92	0.95	1.04	1.07	Close	Yes	Yes
Latin America and the Caribbean	0.99	1.04	0.97	0.98	1.06	1.25	Yes	Yes	Yes
Developing countries	0.86	0.75	0.68	0.99	0.98	1.02	Yes	Yes	Yes

Source: UN (2015a, 2015b).

2.6.3 Goal 3: Promote Gender Equality and Empower Women

One of the sub-targets under this goal is Target 3A: to 'eliminate gender disparity in primary and secondary education by 2005 and in tertiary education by 2015' (Table 2.9). Achievements are a mixed bag. While gender disparity has almost ceased to exist in primary and secondary education, it is still a problem with respect to higher education, except in Asia and Latin America. Among the survey countries it is only Vietnam that is close to target achievement. When it comes to higher education, gender equality is closer to parity in Bolivia and Vietnam compared to the African countries.

2.6.4 Goal 4: Reduce Child Mortality

An important target in the MDGs has been to reduce child mortality. While important progress has been achieved with reducing infant mortality (Table 2.10), there are still serious problems to reduce under-five mortality.

Infant mortality has shrunk considerably in the survey countries. Tanzania has been especially successful in bringing down infant mortality, although it is Vietnam that has the lowest current infant mortality rate. Successful vaccination campaigns to immunize against measles have no doubt contributed to this achievement.

With increasing success in combating infant mortality, attention is

Table 2.10 Infant mortality and one-year-old immunization against measles

	Infant mortality (per 1000 live births)		One-year-old immunization against measles (%)	
	1990	2015	1990	2015
Bolivia	84.6	31.2	53	95
Mozambique	158.0	61.5	59	85
Tanzania	101.3	36.4	80	99
Vietnam	36.5	19.0	88	98

Source: UN (2015b).

Table 2.11 Under-five mortality rates

	Under-five mortality rates (per 1000 live births)		Target achievement
	1990	2015	
Bolivia	123	39	Target achieved
Mozambique	237	87	Close
Tanzania	167	52	Target achieved
Vietnam	51	24	Close
Memo items:			
sub-Saharan Africa	179	86	No
Southeast Asia	71	27	Close
East Asia	53	11	Target achieved
Latin America and the Caribbean	54	17	Target achieved
Developing countries	100	47	No

Source: UN (2015a, 2015b).

gradually turning to reducing child mortality at large. Accordingly, MDG Target 4 has been to 'reduce by two-thirds, between 1990 and 2015, the under-five mortality rate'. The target has been achieved in many developing countries, especially in East Asia and Latin America and the Caribbean (Table 2.11). Both Bolivia and Tanzania have achieved the target, and both Mozambique and Vietnam are close to the target.

Table 2.12 Maternal mortality ratios (per 100,000 live births)

	1990	2015	Target achievement
Bolivia	510	200	No
Mozambique	1300	480	No
Tanzania	910	410	No
Vietnam	140	49	Close
Memo items:			
sub-Saharan Africa	990	510	No
Southeast Asia	320	140	No
East Asia	95	33	Close
Latin America and the Caribbean	130	77	No
Developing countries	430	230	No

Source: UN (2015, 2015b).

2.6.5 Goal 5: Improve Maternal Health

Targets under this goal include Target 5.A: to 'reduce by three quarters, between 1990 and 2015, the maternal mortality ratio', and Target 5.2: '(increase) births attended by skilled health personnel (to 100%)'. The results are shown in Tables 2.12 and 2.13.

Although the maternal mortality ratios have gone down considerably (in the survey countries they have been reduced by over 50 per cent), the results are as a rule far from the ambition set to reduce the mortality ratios by 75 per cent. It is only in East Asia and Vietnam where the results are close to the target (Table 2.12).

Table 2.13 shows the number of births that are attended by skilled personnel. It is only in East Asia where the target of 100 per cent has been reached. But Bolivia has been quite successful and the other countries are not far from the target. In Mozambique and Tanzania the percentage has ameliorated between 1990 and 2015, although at a rather slow rate.

2.6.6 Goal 6: Combat HIV/AIDS, Malaria and Other Diseases

The prevalence of AIDS was very common in the 1980s and 1990s. So it is not surprising that combating AIDS/HIV together with other diseases was an important target among the MDGs, when targets were set in 1990. Target 6.A was thus to 'have halted by 2015 and begun to reverse

Table 2.13 Births attended by skilled health personnel (%)

	1990	2015
Bolivia	47	85
Mozambique	44	54
Tanzania	44	49
Vietnam	77	94
Memo items:		
sub-Saharan Africa	32	52
Southeast Asia	49	82
East Asia	94	100
Latin America and the Caribbean	81	92
Developing countries	57	70

Source: UN (2015a, 2015b).

Table 2.14 Frequency of HIV, malaria and TBC

	HIV frequency % (15–45 living with HIV)		Malaria frequency (notified cases per 100,000 inhabitants)		TBC prevalence rate (per 100,000 inhabitants)	
	1990	2015	1990	2015	1990	2015
Bolivia	0.07	0.11	n.a.	523	400	196
Mozambique	0.70	10.75	n.a.	27,774	791	559
Tanzania	5.19	4.95	n.a.	19,959	334	172
Vietnam	0.19	0.40	n.a.	168	560	209

Source: UN (2015b).

the spread of HIV/AIDS'. Another target was Target 6.C, to 'have halted by 2015 and begun to reverse the incidence of malaria and other major diseases'. Table 2.14 shows some indicators of the frequency of HIV, malaria and tuberculosis (TBC) in the survey countries.

One of the indicators of the frequency of HIV in a country is the prevalence of people in the age cohort 15 to 45 that live with HIV. In 1990 the percentage was below 0.5 per cent in three of the countries, Bolivia, Mozambique and Vietnam, while beginning to be alarmingly high (over 5 per cent) in Tanzania. By 2015 this percentage increased only marginally in Bolivia and Vietnam (still under 0.5 per cent), but continued to be around 5 per cent in Tanzania, and had increased dramatically in Mozambique to almost 11 per cent.

Malaria is a disease that is haunting sub-Saharan Africa, and this goes for Mozambique and Tanzania as well. Although data for 1990 are lacking, the high numbers of detected cases in Mozambique and Tanzania indicate that much remains to be done in fighting malaria: no less than 28 per cent and 20 per cent of the populations of Mozambique and Tanzania, respectively, were infected by malaria in 2015.

The TBC situation is much more promising. In the survey countries, in 1990 the prevalence rate of infected people was less than 1 per cent of the populations. By 2015 the rates had decreased by over half in Bolivia, Tanzania and Vietnam. Only Mozambique has a higher prevalence rate than 0.5 per cent.

2.6.7 Goal 7: Ensure Environmental Sustainability

Goal 7 is a mixed basket with targets that will in the long run contribute to environmental sustainability, not only with respect to, for instance, deforestation and carbon dioxide emissions, but also living conditions for people. We concentrate here on living conditions since data for biodiversity loss not are not readily available for all survey countries.

Target 7.C aims at 'cutting by half, the proportion of people without sustainable access to safe drinking water and basic sanitation' and Target 7.D: 'By 2020, to have achieved a significant improvement in the lives of at least 100 million slum dwellers (worldwide)'. Tables 2.15, 2.16 and 2.17 show some of the results for the survey countries.

The proportion of people using an improved drinking water source increased between 1990 and 2015 in three of the countries: Bolivia, Mozambique and Vietnam (Table 2.15). The increase is especially impressive in the rural areas. The exception is Tanzania – perhaps because of rapidly increasing urbanization – where only 75 per cent of the urban population has access to safe water (compared to 92 per cent in 1990).

The situation regarding improved sanitation facilities is contradictory and somewhat hard to interpret. On the one hand, Tanzania has been most successful, although increasing the percentage with access to improved sanitation facilities from a very low level – from 6 per cent to 31 per cent. On the other hand, considering the levels achieved by 2015, Bolivia and especially Vietnam have been quite successful. Also Mozambique has achieved higher levels than Tanzania (Table 2.16).

One of the biggest problems in the developing countries, as well as in middle income countries, is rapid urbanization. Local communities cannot

Table 2.15 Access to clean drinking water

	Access to clean drinking water (%)			
	Urban		Rural	
	1990	2015	1990	2015
Bolivia	91	97	40	76
Mozambique	72	81	23	37
Tanzania	92	75	45	46
Vietnam	90	99	56	97

Source: UN (2015b).

Table 2.16 Improved sanitation facilities

	Access to improved sanitation facilities (%)			
	Urban		Rural	
	1990	2015	1990	2015
Bolivia	42	61	11	28
Mozambique	34	42	2	10
Tanzania	6	31	7	8
Vietnam	65	94	29	70

Source: UN (2015b).

Table 2.17 Urban population living in slums (%)

	Urban population (million)		Slum population as % of urban population		Slum population (million)	
	1990	2015	1990	2015	1990	2015
Bolivia	3.718	7.000	62	44	2.305	3.080
Mozambique	2.843	8.486	76	80	2.161	6.789
Tanzania	4.830	15.590	77	51	3.719	7.952
Vietnam	13.308	30.722	61	27	8.118	8.295

Source: UN (2015b) and World Bank (2015).

keep pace with infrastructure and housing construction necessary to cope with the massive immigration to the urban areas, especially the already large cities. This means that in spite of making great strides in, for instance, health and education, the housing situation is worsening in most developing countries, with increasing slums. The survey countries are cases in point (Table 2.17).

While the urban population between 1990 and 2015 roughly doubled in Bolivia and Vietnam, it increased by about three times in Mozambique and Tanzania. In the same period the percentage of the urban population living in slums increased in Mozambique from 56 per cent to 80 per cent. In Bolivia, Tanzania and Vietnam the percentage decreased. However, in absolute numbers the slum population increased in all four countries: in Bolivia from 2.3 to 3.1 million, in Mozambique from 2.2 to 6.8 million and in Tanzania it more than doubled from 3.7 to 7.9 million people. In Vietnam the slum population also increased although it decreased as a percentage of the urban population, from 61 to 27 per cent.

2.6.8 Goal 8: Develop a Global Partnership for Development

Under this final goal the MDGs aim for all developing countries to benefit from globalization, for instance, Sub-Goal 8.B, which addresses the special needs of the least developed countries: 'tariff and quota free access for the least developed countries' exports; enhanced programme of debt relief for heavily indebted poor countries (HIPC) and cancellation of official bilateral debt; and more generous ODA [Official Development Assistance] for countries committed to poverty reduction'. Another is Sub-Goal 8.C that addresses 'the special needs of landlocked developing countries and small-island developing States'. This is, for instance, the case of Bolivia.

An overriding goal is for developing countries to be less dependent on international loans and aid. Thus, Sub-Goal 8.D is aims to 'deal comprehensively with the debt problems of developing countries through national and international measures in order to make debt sustainable in the long term'.

Official Development Assistance and debt sustainability

The four survey countries all receive Official Development Assistance (ODA) to a lesser or greater extent. According to the 2014 Organisation for Economic Co-operation and Development Development Aid Committee (DAC) report (OECD, 2014), the two African countries are to a high degree still dependent on aid, in contrast to Bolivia and Vietnam. In 2013 the ODA to these two countries accounted for only around 2 per cent of

Table 2.18 Official Development Assistance and debt service

	ODA as % of GNI		Debt service (as % of exports of goods and services)	
	1990	2013	1990	2015
Bolivia	11.9	2.4	34.2	5.3
Mozambique	42.5	15.1	17.3	4.1
Tanzania	28.5	10.4	31.3	3.0
Vietnam	n.a.	2.5	3.2	1.9

Source: OECD (2014) and UN (2015b).

their respective GNI (Table 2.18). The percentage for Mozambique (15.4 per cent) is quite high even by African standards. Only Liberia, Malawi and Burkina Faso have higher dependency ratios in Africa.

The countries have been haunted by quite substantial debts in the past, with high debt service ratios. This goes especially for the African countries. However, the debt service ratios have dramatically decreased during the last decade, and have now stabilized at around 2–5 per cent.

The information technology revolution

Until the end of the early 1990s, the extension of the fixed telephone lines to cover ever-increasing territory was an important part of the development plans. At that time few could dream of the information technology (IT) revolution that would conquer the world in the coming 25 years, although the MDGs included a vision about what was coming, as expressed in Target 8.F: 'In cooperation with the private sector, make available the benefits of new technologies, especially information and communications'. The indicators were: telephone lines per 100 population; cellular subscribers per 100 population; internet users per 100 inhabitants.

It is clear from Table 2.19 that the extension of fixed telephone lines stagnated, and in some cases even came to a standstill in the 1990s, while the diffusion of mobile telephones exploded. The rapid expansion of cellular telephony – especially to remote areas – has undoubtedly contributed to bringing down death rates in many places. The internet has also expanded rapidly in the developing world, although not as rapidly as mobile phones.

Table 2.19 Fixed telephones, mobile phones and internet users

	Fixed telephones		Mobile phones		Internet users	
	1990	2015	1990	2015	1990	2015
Bolivia	2.7%	8.1%	0.0	96.3%	0.0	39.2%
Mozambique	0.4%	0.3%	0.0	69.7%	0.0	5.9%
Tanzania	0.3%	0.3%	0.0	62.8%	0.0	4.9%
Vietnam	0.14%	6.0%	0.0	147.1%	0.0	48.3%

Source: UN (2015b).

2.7 THE MDGS, INTERNATIONAL MONETARY FUND'S POVERTY STRATEGY AND THE NEW SUSTAINABLE DEVELOMENT GOALS

Some observers have expressed surprise over the apparent lack of impact of the MDGs on the development agenda. For instance, Duncan Green, editor of the Oxfam blog, *From Poverty to Power*, discussed this controversial issue in a blog in July 2015 (Green, 2015). Green asked whether 'MDGs [have] affected developing country policies and spending?' He drew on a study comparing findings from 50 countries by Elham Seyedsayamdost from Columbia University (2014).

She reviews post-2005 national development strategies of 50 countries from 'diverse income groups, geographical locations, human development tiers and ODA levels to assess the extent to which national plans have tailored the MDGs to their local contexts'. Of the 50 development plans under review, 32 have adopted the MDGs as planning and development tools. Some have 'localized' them by using various adaptation strategies, changing the target date or setting additional goals or sub-goals, in order to make targets and indicators more relevant for sub-national planning. 'A high correlation is detected between income groups, PRSP (Poverty Reduction Strategy Papers) status and ODA dependency of countries and their propensity to incorporate the MDGs in their planning instruments.' The PRSP are documents prepared by the member countries through a participating process involving domestic stakeholders as well as international development partners such as the International Monetary Fund (IMF) and the World Bank. The PRSP are documents required by the IMF and the World Bank before a country can be considered for debt relief with the Heavily Indebted Poor Countries (HIPC) initiative (see IMF, 2001, 2004, 2011a, 2011b).

The correlation with income level and MDG alignment is understandable. Why, for instance, should a middle income country decrease its percentage level of say people living on less than 1.25 USD per day, by 50 per cent, if it is already in the lower range (example: Bolivia in Table 2.6)? With respect to the IMF and the PRSR process, 'a closer examination of the national plans indicates that PRSP countries are more likely to have aligned their national plans with the MDGs'. However, the countries that have not aligned their plans with the MDGs all belong to the middle income countries and they are also least dependent on ODA. The same with the HDI level and proclivity to adopt or adapt the MDGs: the higher the HDI, the less likelihood that the national plans have been aligned with the MDGs. 'More disconcerting is the lack of connection between planning and implementation . . . (I)n fact, countries that have not integrated MDGs into their national plans are just as likely to allocate government funds to the social sectors as the MDG aligners'. It thus seems that even if the MDGs are reflected in national plans, they apparently have little or no influence on how the government spends its money (reflecting the decisive role of the minister of finance who often overrules decisions by the minister of planning).

Sustainable Development Goals (SDGs)[1] will follow in the MDGs' footsteps. SDG is an agreement by UN member states to develop sustainable development goals, which will build on the MDGs and 'will converge with the UN development agenda' (UN, 2015b). They will hopefully be better coordinated with national development plans than has been the case with the MDGs.

To sum up, Bolivia and Vietnam have done relatively better than Mozambique and Tanzania in achieving the MDGs. On the other hand, Mozambique and Tanzania have been more successful than average for sub-Saharan Africa.

NOTE

1. It is proposed (as of August 2015) that the SDGs will include 17 goals and 169 targets. The goals are: 1. End poverty; 2. End hunger; 3. Well-being; 4. Quality education; 5. Gender equality; 6. Water and sanitation for all; 7. Affordable and sustainable energy; 8. Decent work for all; 9. Technology to benefit all; 10. Reduce inequality; 11. Safe cities and communities; 12. Responsible consumption by all; 13. Stop climate change; 14. Protect the ocean; 15. Take care of the earth; 16. Live in peace; and 17. Mechanisms and partnerships to reach the goals.

REFERENCES

AFDB (2015), *African Statistical Yearbook 2014*, available at http://www.afdb.org/en/knowledge/publications.

ECLAC (2015), *ECLAC Presents its Overview of the Region's Economies in 2015 and its Projections for Next Year*, Santiago de Chile, available at http://www.cepal.org/en/news/eclac-presents-its-overview-regions-economies-2015-and-its-projections-next-year (accessed 7 December 2015).

FAO (2015), *Definitions of Undernourishment*, available at http://mdgs.un.org/unsd/mdg/Metadata.

GPO (2014), *Statistical Yearbook of Vietnam 2013*, Hanoi: Government Printing Office.

Green, D. (2015), 'Have the MDGs affected developing country policies and spending?, available at http://www.oxfamblogs.org/fp2p (accessed 24 July 2015).

IMF (2001), *Bolivia: Poverty Reduction Strategy Paper*, available at http://www.imf.org/external/prsp.

IMF (2004), *Vietnam: Poverty Reduction Strategy Paper*, available at http://www.imf.org/external/pubs.

IMF (2011a), *Mozambique: Poverty Reduction Strategy Paper*, available at http://www.imf.org/external/pubs.

IMF (2011b), *Tanzania: Poverty Reduction Strategy Paper*, available at http://www.imf.org/external/pubs.

OECD (2014), *Official Development Assistance 2013*, available at http://www.oecd.org/dac/stats/data.htm.

Paus, E. (2013), 'Getting development right', in E. Paus (ed.), *Getting Development Right*, New York: Palgrave Macmillan, pp. 11–35.

PWT (2015), Penn World Tables, available at http://pwt.sas.upen.edu/php-site/pwt71.

Seyedsayamdost, E. (2014), 'MDG-based national planning: an assessment', available at http:// www.oxfamblog.org/fp2p/MDG-Assessment_ES.pdf.

UIS (2015), UNESCO Institute of Statistics, available at http://www.uis.unesco.org.

UN (2015a), *The Millennium Development Goals Report 2015*, New York: United Nations.

UN (2015b), *The UN Official MDG Monitor*, available at http://mdgs.un.org/unsd/mdg.

UNDP (2014), *Tanzania Human Development Report 2014: Economic Transformation for Human Development*, available at http://www.hdr.undp.org/sites/files/thdr.

UNDP (2015), *Human Development Report 2015*, available at http://www.hdr.undp.org/.

World Bank (2015), *World Development Indicators*, available at http://wdi.worldbank.org/tables.

3. Emerging innovation systems in developing countries: Bolivia, Vietnam, Tanzania and Mozambique*

Claes Brundenius, Carlos Aguirre-Bastos, Tran Ngoc Ca, Bitrina Diyamett and Maximiano Dgedge

> The national innovation system is an open, evolving and complex system that encompasses relationships within and between organizations, institutions and socio-economic structures which determine the rate and direction of innovation and competence-building emanating from science and experience-based learning. (Lundvall et al., 2009b, p. 6)

3.1 THE ORIGINS OF THE NATIONAL INNOVATION SYSTEM APPROACH

The concept of the national innovation system has become a powerful tool in understanding and analysing the role of science, technology and innovation (STI) in growth and development. This chapter traces the roots of the concept of the national innovation system and describes how it has emerged in the four case countries, Bolivia, Vietnam, Tanzania and Mozambique.

3.1.1 A Growing Interest in Science and Technology Policies

Soon after the Second World War there was an emerging interest in the West about the role of science and science policies in economic growth and development. The main reason was the impact of basic scientific research, and especially so-called big science as witnessed during the war. The most important example is of course the Manhattan project that, thanks to the concerted efforts by thousands of scientists, managed not only to develop a devastating weapon – the atomic bomb – but also later tamed the atom into peaceful nuclear energy. The role of government support was now seen as crucial in science and technological strategies. The post-war competition between the new superpowers, the United States and the Soviet Union, is a case in point, with massive scientific and technological input that had to be concentrated in the arms race and race to the moon.

3.1.2 The Need to Measure Research Inputs

It soon became clear to policy makers that some measuring tools were needed in order to plan investment in science and technology capacity building. Although work and collection of data on science and technology had already started at the National Science Foundation in Washington, DC in the 1950s, it was the Organisation for Economic Co-operation and Development (OECD), with its Secretariat in Paris, that took the lead in developing indicators that could measure primarily science and technology inputs in general, and research input in particular. Later, it became evident that it was also important to measure research output. These indicators divided research into the three components: basic research, applied research and experimental development. The indicators estimating inputs into research and experimental development relate to both expenditures and research personnel. The first rudimentary guidelines were adopted at a meeting in Frascati (Italy) in 1962 and the aim was that these guidelines would become a manual to be used by all OECD member countries. This manual has been continually updated as the so-called Frascati Manual, with the latest version published in 2015 (OECD, 2015). The Frascati Manual is now a standardized and uniform way of measuring research and development (R&D) worldwide.

3.1.3 The Black Box and the Residual

Meanwhile, in the 1960s there was emerging interest by economists to look into what was called the 'black box' (Rosenberg, 1982). Economists were increasingly becoming aware that the traditional production function was not sufficient to explain economic growth (Y). In addition to Capital (K) and Labour (L), there was another factor of production obviously missing. In the 1920s Cobb-Douglas had tried to estimate the so-called third factor, or 'residual factor', in economic growth, a factor that in addition to capital and labour would explain growth and why growth rates differ. But it was Robert Solow, a future Nobel laureate, who in 1957 became a pioneer of groundbreaking work with his seminal paper, 'Technical change and the aggregate production function'. Solow estimated that the 'residual' – roughly speaking – could account for as much as 60 per cent of growth.

No wonder that interest in the role of science and technology increased in the 1960s and 1970s. Many institutes specializing in the 'study of science' were born. Examples are the Science and Policy Research Unit (SPRU) at Sussex University in the UK and the Research Policy Institute (RPI) at Lund University in Sweden, both starting research programmes in 1966.

3.1.4 The Concept of National Innovation Systems

In the 1980s the concept of the national innovation system was born. The main contributors were Chris Freeman, Director and founder of SPRU, and Bengt-Åke Lundvall from Aalborg University in Denmark. Freeman used the concept in 1982 (although without indulging in any definitions) and he thus used the concept in the context of his interest in international trade and competitiveness. Nelson later defined and revised the concept of innovation systems (Nelson, 1993). Lundvall and his colleagues at Aalborg started at the micro level, where innovation is seen as a result of the interaction between user and producer (Lundvall, 1985, 1992). Both scholars were critical of the simplistic linear relationships tacitly assumed in STI studies at the time (Godin, 2010). That is, that there is a direct line from basic to applied research, from applied to experimental development, to innovation and its commercial application and diffusion on the market (OECD, 1997).

Both approaches stressed the importance of State intervention, with infant protection needed in 'catch-up countries', a policy not much in vogue in the neo-liberal political environment of the 1980s.

3.1.5 Science and Technology – an Exogenous Factor?

In neo-classical economic theory the technology factor had been considered an exogenous factor – decided outside the economic sphere – and having an impact (positive or negative) just like the weather or natural disasters. With Solow and other economists (econometricians primarily) there was simultaneously a growing interest in growth accounting, where the 'residual factor' was embedded as Total Factor Productivity (TFP), together with the traditional factors of production, capital investment and labour inputs.

3.1.6 Schumpeter and Innovation

Innovation system scholars are generally influenced by Joseph Schumpeter, who in the 1930s stressed the impact of innovations, and especially the entrepreneurial innovator, who was vital for growth and development

(Schumpeter, 1943). Richard Nelson and Sidney Winter in their pioneering work on evolutionary economics strongly criticized the traditional view of seeing innovation as a linear process from invention (through research and development) to patenting and its final commercialization (Nelson and Winter, 1982). Freeman, Nelson and Lundvall among others in the so-called evolutionary school stressed the different roles of actors in the innovation system, whereby their interaction was considered crucial for success.

3.1.7 Knowledge-based Economy or Learning Economy?

The learning process was underlined as one of the most important aspects of the interaction, not least learning by the users who after all are the main purpose of the innovations. The 'knowledge-based economy' became a new catchword in international organizations, such as the OECD, the World Bank and the European Union (EU). However, Lundvall and others claimed that the 'learning economy' was a more dynamic description of the process.

It became clear that the Frascati Manual did not capture all the elements of 'innovation' and thus there emerged a need for a manual on measuring innovation activities, other than research in laboratories. The first edition of the so-called Oslo Manual was published by the OECD in 1992, with the title, *Guidelines for Collecting and Interpreting Innovation Data*. In Latin America the Oslo Manual was criticized for not taking into account 'the innovation landscape' in developing countries, and Red de Información sobre Ciencia y Tecnología (RICYT), a monitoring science and technology observatory, published its own version of the manual, the so-called Bogotá Manual. The Bogotá Manual was subsequently incorporated into the third edition (2005) of the Oslo Manual in an annex entitled, 'Innovation surveys in developing countries'.

3.1.8 The Role of Universities in Innovation Systems

There are many actors in an innovation system and lately the focus has been on the role of universities and other academic institutions. Universities have for too long been looked upon as 'ivory towers' without contact with the real world. This conception is rapidly changing as universities not only become relevant for capacity building by lifting the learning levels of countries but also becoming involved in research that is relevant for growth and development. In many countries – not least in the South – universities have become involved in extension projects and other 'third missions' (after education and research) and many see the developmental university as a future model to address the problems of development and social justice (Brundenius et al., 2009; Göransson and Brundenius, 2011).

3.1.9 Measuring STI in Africa

The Consolidated Plan of Action (CPA) was adopted by the African Ministerial Council on Science and Technology (AMCOST) in 2005, with a specific programme on the development and use of STI indicators. In 2007 the ASTII initiative was set up by the New Partnership for Africa's Development (NEPAD) as a means to support the CPA to improve conditions for STI policy analysis and planning. The ASTII programme has so far produced two editions (NEPAD, 2011, 2014) of the *African Innovation Outlook* with STI statistics and analyses for many of the African Union (AU) member countries.

In 2014 the AU launched a new long-term strategy for Science, Technology and Innovation for Africa until 2024 (STISA).

3.1.10 The Institutionalization of National Innovation Systems

In 2001 Sweden became the first country in the world to institutionalize a national innovation system by delegating a specific agency to deal with the 'Swedish Innovation System': VINNOVA (Swedish Governmental Agency for Innovation Systems). This idea was to a large extent influenced by the OECD and also by the research community in Sweden and the Nordic countries (Eklund, 2007).

The Swedish experience has been followed in other European countries, although without such an explicit reference to innovations systems as such. There is also a growing interest in institutionalizing the innovation system concept in developing countries. A case in point is InnovaChile in Chile, which to a large extent is influenced by the Swedish idea.

3.1.11 The Role of Indicators in Science Technology and Innovation Policies

There is a growing awareness in most parts of the world that STI plays an important – perhaps crucial – role in development strategies. But in order to plan and formulate STI policies it goes without saying that some kind of benchmarking is necessary. Interesting books have been written on this subject lately, such as Gault (2013).

Gross Expenditures on Research & Development (GERD) has become such a benchmark. GERD, as percentage of gross domestic product (GDP) per capita, is used as a measure of the R&D intensity of a country, and has become a policy target as important as economic growth itself. One of the targets in the *Europe 2020 Strategy* is that the EU member countries should invest 3 per cent of GDP in R&D. Such targets are also often seen in many countries outside Europe. AMCOST at its meeting in 2005 established the goal that the AU members should reach a GERD of 1 per cent of GDP by the end of the decade (that is, around 2010). However, this target has not been reached (not even by South Africa), and the STISA plan is much more cautious about setting too ambitious targets for 2024.

3.1.12 The Research System in the Four Survey Countries by Numbers: A Snapshot

Below we illustrate the research systems in Bolivia, Mozambique, Tanzania and Vietnam by comparing some basic indicators, in comparison with some other countries. The data are collected in most countries in the world – as a rule annually – and in principle the data are standardized using the Frascati Manual definitions. The data are thus – in principle – comparable but some caution has to be used in interpreting the data. This goes especially for countries that are newcomers in this field (of collecting data on R&D), and for most countries in Africa, but also Latin America (for example, Bolivia) and Asia (for example, Vietnam). But with this caution expressed one may still draw some interesting conclusions from the data presented in Table 3.1.

The overall investment in R&D is the GERD and is usually expressed in its relation – percentage – to the country's GDP. While the GERD percentage is over 2 per cent in many OECD countries (here represented by Sweden with 3.41 per cent) it is considerably lower in most developing countries, with China as an impressive exception (1.98 per cent). When we look at the survey countries there are some apparent anomalies. The two countries with the highest GDP per capita – Bolivia and Vietnam – are the countries that spend the least on R&D as a share of GDP, while one would expect the contrary: the higher GDP per capita, the higher R&D intensity. On the other hand, Vietnam has by far the highest researcher density – 1170 researchers per million inhabitants. This is a very high figure by international comparison: it is almost at the level of Brazil and China, 5 times higher than Bolivia and 16 times higher than Tanzania. But once again it should be stressed that definitions could vary, and particularly definitions of a 'researcher' can vary in country surveys, even if there are quite clear and unequivocal definitions in the Frascati Manual.

Women do research to a high degree in all four countries, especially in Bolivia (67.5 per cent) but also in Vietnam where female researchers account for over 40 per cent of the total.

Table 3.1 R&D indicators in the survey countries compared with other selected countries

	Bolivia 2010	Mozambique 2011	Tanzania 2010	Vietnam 2011	Sweden 2011	Brazil 2010/12	South Africa 2010	China 2011	Cuba 2012
GERD million PPP	70.8	89.2	290.5	703.4	11,324.8	24,451.0	3623.2	213,444.3	563.1
GERD PPP/Cap	6.93	3.85	7.19	7.92	1,190.7	124.2	70.4	154.8	51.19
GERD % of GDP	0.15	0.46	0.52	0.19	3.41	1.21	0.76	1.98	0.42
BERD % of GERD	6.0	–	–	26.3	71.1	39.8	57.7	72.8	–
GOVERD % GERD	58.9	54.3	13.4	57.9	6.0	21.4	21.1	18.5	100.0
HERD % of GERD	30.6	40.0	86.5	15.8	21.0	38.8	18.9	8.6	–
External financing	1.9%	78.1%	42.0%	4.0%	11.1%	n.a.	12.1	1.0%	5.0%
R&D personnel (HC)	3450	3313	5788	134,781	125,130	469,257	55,531	4,617,210	14,859
Researchers (headcount)	2153	1588	3102	105,230	80,039	234,797	37,901	1.905,899	4655
Researchers per million inhabitants	212	142	74	1170	8471	1203	737	1393	413
Female researchers	67.5%	32.2%	25.4%	41.7%	37.2%	n.a.	41.7%	n.a.	48.4%
GERD/researcher 000 PPP	47.9	56.2	93.6	6.7	139.4	97.2	95.6	96.1	121.0

Note: BERD – business enterprise research & development; GOVERD – government research & development; HERD – higher education research & development; PPP – purchasing power parity.

Source: UIS (2015).

With respect to performance of R&D, there are some interesting differences between the countries. Government institutes in Bolivia, Mozambique and Vietnam account for over 50 per cent of all research carried out in these countries. In contrast, the corresponding figure in Tanzania is only 15 per cent with most of the research performed at universities. Interestingly, the role of business enterprises is quite large in Vietnam (26 per cent).

Research is to large extent financed by foreign donors in Mozambique (78 per cent) and Tanzania (42 per cent), while dependency on foreign sources is quite low in Bolivia (1.9 per cent) and Vietnam (4.0 per cent).

After this snapshot, the following four sections present an account of the emergence of the national innovation system approach in each of the four case countries, and how this has manifested itself over the last six decades.

3.2 EMERGENCE OF THE NATIONAL INNOVATION SYSTEM APPROACH IN BOLIVIA

3.2.1 The National Innovation System Approach in Latin America

Science and technology (S&T) policy has been subject to analyses and debate in Latin America since the early 1950s. In 1960, the recently created regional science office of the United Nations Educational, Scientific and Cultural Organization (UNESCO) convened, in Caracas, Venezuela, the first meeting of responsible officers for science policy in the region and initiated efforts to promote the creation of national research councils. In 1965, in Santiago de Chile, the First Conference on the Application of Science and Technology for the Development of Latin America was held as the initial step of a 'Permanent Conference' that would take place at the VI Conference in Brasilia, Brazil in 1985 (UNESCO, 2010). Further, UNESCO promoted extensive regional debates on science policy up to the 1990s when the concept of innovation was introduced in the STI discourse of most countries.

Policy discussed in the earlier period was conducted to a large extent by the scientific community. It was clearly a 'science push' effort following a linear S&T development model aimed at strengthening research organizations, universities and research councils. During this period governments responded mainly to appeals emerging from the scientific communities. UNESCO can rightly claim to have been the key instrument for a 'science push' model in the region, with considerable effects.

In the mid 1950s and following the nascent economic model of import substitution being strongly promoted by the United Nations Commission for Latin America (ECLAC) (Prebisch, 1949), and with little interaction with the scientific community, a new policy movement was born in the region that emphasized endogenous technology development as a way to economic growth. Martinez Vidal and Mari (2002) call this movement the 'Latin American School of Thought in Science and Technology' (the School). This is a time when the demand-oriented model began to be introduced in the region. It is also a time when the innovation system approach to S&T began to take shape.

The School was clearly influenced by the import substitution model that considered that the economy of periphery nations had excessively specialized in the production and export of commodities and that this specialization was deteriorating the terms of exchange and restricting foreign trade. It was considered that such specialization had also resulted in a dependence on imported capital goods, other critical inputs and foreign technology, thus leading to structural underdevelopment. As a consequence, in the 1950s, the region saw the initiation of more active policies of industrialization as a preferred way to emerge from such underdevelopment.

One of the early products of the School was the so-called 'Sabato's Triangle' proposed by Sabato in 1967 (Sabato, 1964, 1975; Sabato and Botana, 1968). The main idea of the proposition was to understand and construct a vigorous S&T infrastructure linked to the productive structure of countries. As such, the vertices of the triangle comprised government, productive structure and the S&T infrastructure. Sabato put forward a set of different relationships among the elements of the triangle: intra-relations (within each vertex); inter-relations (within the vertices); and extra-relations (with the government directing the other two vertices). This proposition could certainly be understood only in terms of a system.[1]

At the same time, the Organization of American States (OAS) began developing a different S&T policy concept from the 'science push' model and actively and successfully supported operational activities in its application. The S&T concept developed in the OAS followed mainly the thinking of Maximo Halty (1979). Sagasti and Hodara (1981) have fully analysed and described Halty's approach, concluding that his emphasis on technological development, although still following a 'supply side' vision, amply influenced the thoughts of the School and future OAS programmes by proposing a model of technological learning, based on promotional measures, governmental protection of early local technology efforts, control of technology transfer and an administered system of incentives and prices. It is considered by the authors that Halty's work constituted a first step towards the adoption of the innovation system approach to S&T development in the region.

In 1964 a Consultative Committee of OAS called for the need of 'establishing a state policy in science and technology, and establishing a pre investment system that would guide and receive the benefit of the possible knowledge transfer that is imported through technical aid'. This call gave birth to an extensive effort of OAS that reflected the thoughts of the School, made operative by the adoption of the 'Regional Program of Scientific and Technological Development', approved by the Heads of State in 1967 in Punta del Este, Uruguay and made operational in February 1968, through the Maracay Resolution. The conceptual and operative approach showed a clear differentiation between purely guided scientific research and endogenous technology development policies, which emphasized the closing of the technology gap and not the promises of science. The new approach did include, however, the recognition of the importance of research.

At the end, what prevailed at OAS is what Seidel (1974) called a 'hybrid model', in which the conceptual approaches of the School and the proposals set forth by the scientific community – forcefully led by Argentinean Nobel Prize winner Bernardo Houssay – to support scientific infrastructure were integrated. As a result, during the first ten years of the programme a very large number of research institutions were created or received financial support in the region.

It is also in this period that several other organizations participated in the efforts to further develop the conceptual and operational modes developed by the School. At the multilateral level the United Nations Conference on Trade and Development (UNCTAD) and later the United Nations Industrial Development Organization (UNIDO) introduced the new thinking in their own approaches and programmes. At the bilateral level, the International Development Research Centre (IDRC) of Canada and the Swedish Agency for Reseach Cooperation (SAREC) become key actors, with some other minor participants from agencies in developed countries. The Inter-American Development Bank became a very strong player as well, and is today not only the largest donor of cooperation in STI in the region, but also one of its key 'think tanks'.

Both ECLAC and UNCTAD strongly advised in the mid 1960s on the need to promote technology development and in the latter case UNCTAD I defined a resolution regarding the transfer of technology, which gave rise to an extensive negotiation on what was later called the International Code of Conduct for the Transfer of Technology or the Code for the International Transfer of Technology, depending on whose side the problem of transfer was viewed. The above resolution also called for a review of the patent system and issues of proprietary technology. Latin American thought is clearly reflected in the discussions that were conducted during the negotiations of the Code, which in the end was never adopted (Aguirre-Bastos, 2001).

Although the contribution to the School's thinking and its influence can be perceived from many angles and individuals, one important example can be recognized in the definition of the Andean Pact technology policy (Jaramillo and Aguirre-Bastos, 1989). In the early 1970s the Junta del Acuerdo de Cartagena (the Technical Secretariat of the Andean Pact) led a new conceptual movement to differentiate science from technology in policy, objectives and practices and to characterize the overall effects of technology imports. The studies conducted by the Junta were pivotal for understanding issues related to technology and its impact on production, particularly industry. Valuable empirical information was collected and analysed on the relationship between technology and production in an economic integration context, based on the concept of technology as an ingredient of production and therefore functional to the definition of economic strategies.

The technology policy adopted in the early 1970s and applied throughout the 1980s was designed to take advantage of the demands generated by the economic integration process in order to open up the possibilities of linking technology to the construction of an open market, subordinate foreign investment and technology transfer to goals of communitarian interest and bringing technology to the production process.

The creation of an extended market had implications that could be channelled to technology issues. Under the State's guidance, national public and private enterprises were favoured; thus it was foreseen that any investment and technology transfer process should provide advantages to the national enterprise with respect to its participation in that market. The new concepts of technology born in the sub-region also resulted from analysis of the characteristics of technology trade, such as, for example, licence contracts and the distortions created by 'turnkey' agreements. The analysis led to the understanding of the relations between economic development, dependence and integration, and contributed to typify better the nature of technology issues involved in an import substitution model and the processes for the international search of technology.

Within this context, the Andean Pact adopted Decisions 24 (on foreign investment and technology transfer), 85 (on industrial property rights) and 84 (a common technology policy) to regulate the flow of technology and promote endogenous development. The underlying studies provided the basic building blocks of a 'defensive and control school' that was looked upon as an almost exclusive ingredient of a technology development strategy. It would take some more years for national and regional organizations to become involved in issues such as innovation and technical change, especially at the influence of the School's relation with the Development Centre of the OECD, led at the time by Nicolas Jequier. The innovation concepts were further developed in the region by those scholars that were receiving higher education at SPRU, University of Sussex.

Sagasti (1983) made an extensive review of the early national innovation system approach to S&T policy in the Latin American region. He recognized that the efforts to deal with different issues in S&T policy, as briefly discussed above, were mostly an attempt to order in a systematic way the different issues, factors and institutions that intervene in the S&T development of the region. Sagasti considers that the formalization of the innovation system approach in the region took place between 1971 and 1972, through four main works that were representative of the ongoing thinking of the time: (1) the essay by Maximo Halty on the 'production, transfer and adaptation of industrial technology'; (2) the work of Alberto Araoz on 'An approach to Science and Technology Policy and Planning'; (3) the work of Felix Moreno on a 'Model for a system of technology production, selection and transference'; and the doctoral thesis of Sagasti, 'Towards a methodology for planning science and technology in undeveloped countries'.

These approaches share several innovation system characteristics. On the one hand, the role of the demand for knowledge and technology as they can respond to national objectives; in addition, the approaches highlight the importance of the import and transfer of technology that need to be balanced with the local production of technology, all calling for the development of local capabilities, for which the intervention of the State is crucial, and recognize that explicit S&T policies need to be defined. As also pointed out by Sagasti, Halty proposed the concept of a technology strategy, while Araoz introduced the idea of critical areas for decision making as mechanisms to systematize the formulation of policies. Moreno provided for the institutional and normative reordering of policy and Sagasti in his thesis highlighted the importance of the policy and planning process, the harmonization of explicit and implicit policies and the different decisional categories for planning scientific and technological development.

In the case of the Andean Pact, its early studies and the need to involve the scientific community in the economic integration effort led to the definition of the Andean System for Science and Technology, conceived as a permanent organism for coordination, advice, financing and study (Jaramillo and Aguirre-Bastos, 1989). Other proposals for a national innovation system approach to S&T, and later, innovation, can be found in this early period. The efforts of the Andean Inter University Centre (at present Inter University Centre for Development, CINDA) is particularly recognized, as described next, as it served the emergence of the national innovation system approach in the case of Bolivia in the mid 1970s.

Table 3.2 Periods of creation of the first national S&T organizations in Latin America

1945–59	1960–73	1974–87	1988–2001
Argentina (1950) Mexico (1950) Brazil (1951)	Bolivia (1960) Jamaica (1960) Uruguay (1961) El Salvador (1962) Cuba (1962) Paraguay (1963) Chile (1967) Venezuela (1967) Colombia (1968) Peru (1968) Trinidad & Tobago (1968) Costa Rica (1972) Guyana (1972)	Guatemala (1974) Dominican Republic (1974) Barbados (1977) Nicaragua (1977) Ecuador (1979) Dominica (1980) Granada (1982)	Panama (1990) Honduras (1993)

Source: Adapted from www.unesco.org/new/en/natural-sciences/science-technology/sti-policy/a-brief-history-of-unescos-science-policy-programme/ (accessed 15 November 2015).

3.2.2 Emergence of the National Innovation System Approach in Bolivia

As discussed above, during the 1950s and 1960s, the Latin American countries initiated the process of institutionalizing the development of S&T at the functional levels of the State. Table 3.2 shows how this process evolved over time. Bolivia was one of the first countries in the region to establish a policy-making body. But, different from all other countries, the government created the National Academy of Sciences as an autonomous public organization in charge of the formulation of science policy (Aguirre-Bastos, 2002). To define technology development policies, the Ministry of Industry created the National Directorate for Standards and Technology also in the 1960s.

Considering the need to define an institutional scheme that would better respond to the requirements of social and economic development, and considering the ongoing discussions on S&T policy in the region, Bolivia decided to make operational the innovation system approach, and as such, was the first country in the region to formally do so, even though the approach had already been theoretically discussed extensively in other countries.

In 1977 the Bolivian government adopted Decree 15111 establishing the National System for the Development of Science and Technology with a National Council and a Directorate in the Ministry of Planning as its executive body. Carlos Aguirre-Bastos, who had participated in the development of the 'Latin American School', was charged with leading the nascent Directorate.

The birth of the new Directorate was not problem-free. The National Academy of Sciences and the Directorate for Standards and Technology considered that their attributes had been infringed. The universities also felt that they had been put aside in the new institutional arrangement. In spite of the fact that it was a Member of the Academy who had been appointed as the first director, it took some time to overcome such worries and establish an environment of trust and collaboration so that the innovation system could operate well.

In 1980, the Directorate defined an extensive work plan to consolidate the leadership of the innovation system. By 1981 the government approved a set of policy and strategic guidelines to strengthen it. Both the early design of the institutional arrangement and later the policy design and strategic guidelines were strongly influenced by the system approach taken by CINDA (1977).[2]

The model is schematized in Figure 3.1. In it, the system is composed of four strongly articulated sub-systems, following specific functions:

a. Incorporation of knowledge comprising those organizations (and functions) that produce new knowledge.
b. Intermediaries of knowledge comprising those organizations (and functions) that adapt foreign knowledge or become the link between knowledge and production.

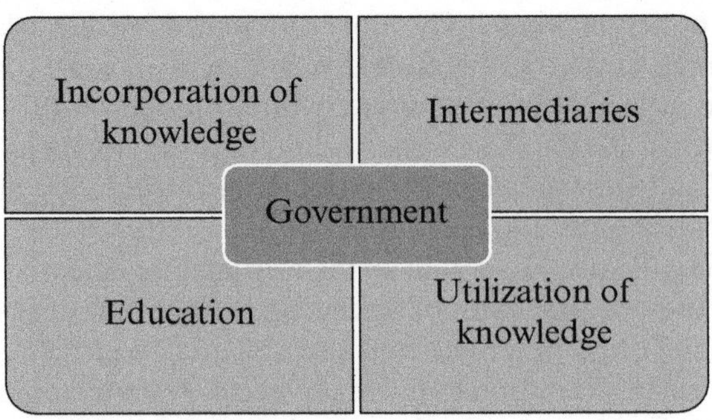

Source: Adapted from CINDA (1977).

Figure 3.1 The science and technology development system

c. Utilization of knowledge comprising those organizations (and functions) that constitute the demand and use of knowledge.
d. Education comprising those organizations that educate and train human resources for the system.

The system is controlled and regulated by a higher body, called the Government System, which is not necessarily only a public body but also refers to the governance mechanisms of the system. The flow of knowledge constitutes the essence of the system, and this flow not only involves a flow of resources but also includes the existence of a market, and the traded good is technology or technology-based innovation.

In the application of the system approach, the Directorate for Science and Technology invited the Bolivian scientific community and the productive sector to prepare the national S&T development plan on the basis of the approved guidelines. It was concluded in 1982 but because of the prevailing political and economic situation of the country it was never formally adopted. In spite of this, however, the guidelines served as a reference for later planning processes, especially in the university system.

In 1991, considering that the Directorate had lost its influence in the sphere of the Ministry of Planning, the government created the National Council for S&T and the Executive Secretariat under the Vice President of the Republic, through Decree 22908 of 23 September. This new legal instrument again recognized the existence of the system. Later in 1995 after a set of demand and supply studies ordered by the vice president, a short-term policy and action plan were adopted. The important role played by two vice presidents in the strengthening of the system, Luis Ossio Sanjines and Victor Hugo Cardenas, should be acknowledged.

The short-term policy and action plan had as a main objective the generation of an appropriate context for the development of STI that would support sustainable development goals. This was the first time in a policy statement that the concept of innovation was introduced formally. Another key objective of the policy and action plan called for a strengthening of the system, a goal that was successfully reached by establishing strong institutions for technology development and information and communications technology (ICT) applications. The important role of OAS and UNIDO is recognized here, as both organizations supported the application of the short-term policy and action plan.

Although the institutional framework went through new changes after 1997 and later in 2006, the system approach initiated in 1977 was maintained and enriched. In 2006 the National Development Plan and in 2009 the National S&T Development Plan call for the operation of the national system of innovation and the national system for science, technology and innovation, respectively.

3.3 THE EMERGENCE OF THE INNOVATION SYSTEM IN VIETNAM AND THE INSTITUTIONAL SETTING

3.3.1 The Institutional Framework and Actors of Changes

Although the establishment of Vietnam S&T organizations, and later on the innovation system, dates back to the mid 1960s, the subject of management of S&T was brought to Vietnam at the end of the 1970s and beginning of the 1980s. The Unit for S&T Policy Research was created in 1978 in the then State Commission for Science and Technology. Gradually, it evolved into various forms and became the Institute for Science and Technology Management (ISM) in 1986. In 1996, it merged with the Institute for Strategy and Forecasting of S&T (also under the same Ministry of Science and Technology) and became the National Institute for S&T Policy and Strategy Studies (NISTPASS). As a leading think tank under the Ministry of Science and Technology in STI policy, ISM and then NISTPASS were at the forefront in promoting the systemic approach of the innovation system in Vietnam (Vu Cao Dam, 2008). During the 1990s, other think tanks such as the Central Institute for Economic Management (CIEM) and the Institute for Policy and Strategy for Agriculture and Rural Development (IPSARD) also began to pay more attention to the subject.

3.3.2 Changes of Policy Thinking and Practices for Science, Technology and Innovation in Vietnam

If one looks at the emergence of the innovation system, it is useful to examine the process of changes in innovation policy thinking and the implementation of policies in STI practices. Via this process, the notion of innovation is revealed as having evolved over time, and not having been introduced at a particular time.

The first reforms efforts: pre-Renovation (before the mid 1980s)

By the late 1970s, Vietnam experienced a dire socio-economic situation that lasted into the mid 1980s. This was associated with the international isolation imposed by the West together with China and other Asian regional countries after Vietnam's involvement in the process against the Khmer Rouge regime in Cambodia. This put additional hardship on

Vietnam after the country had already been heavily hurt by a protracted war. After reunification in 1976, economic changes in the South did not bring the desired results. Companies (mainly State owned) suffered heavy losses caused by bad macroeconomic management while agriculture could not produce sufficient food. All of this led to self-initiated reforms by individuals and some 'underground' organizations, despite the rigid dogmatic rules of management. They tried to find new ways of production and to manage their resources and income. This was known as the *pha rao* ('fence-breaking') reform. A 'three plans' mode of production was introduced: as well as State-regulated plans, there were now additional plans to take care of the needs of the enterprises and individuals themselves.

Up to this period, all S&T activities had been under the control and monopoly of the State. No private sector activities were encouraged, and R&D, production and educational activities were separate activities. In 1981, Decision 175, introduced by the then ISM (the first and only think tank working on S&T), which allowed economic contracts between research organizations, can be seen as the precursor of the series of reforms that would follow in the next decades.

Following Decision 175, a series of reform policy documents were introduced including Decision 134 in 1987 (complemented by Decree 35 in 1992) to allow R&D organizations to enter into contractual relationships with individuals and non-State organizations and Decision 268 in 1990 to allow R&D organizations to do outside research business. Regarding foreign direct investment and technology issues, the Ordinance of Foreign Technology Transfer in 1988 and on Intellectual Property Rights (1989) are among the most notable changes. These can be seen as early stages of S&T policy development. However, these efforts were still in the traditional mode of S&T thinking (more of a R&D focus) and with little or no mention of innovation.

New impetus for further changes: the Renovation (*doi-moi*) era after 1987

Initial reforms did not lead to radical changes and poor macroeconomic management decisions led to financial crises, including that of hyperinflation and severe food shortage in 1986. This created the momentum to produce the most radical reform at the Sixth Party Congress, also known as the Renovation *(doi-moi)* reform. This concerned issues such as relaxing management of foreign direct investment (FDI), land, foreign trade, banking, State-owned enterprises (SOEs), private and household enterprises and the agricultural sector. Moreover, a series of important decisions in land management, banking, pricing, investment and SOEs were introduced.

During this period, the Soviet-style management of S&T activities was gradually changed and a new legal framework for S&T-based development came into force (Tran Ngoc Ca, 1999). In addition to the first few legal documents related to FDI and technology transfer, macro targets for S&T were set by Party resolution with 2 per cent of the State budget earmarked for S&T. The mid 1990s also witnessed the creation of a number of new organizations in S&T such as the National Centre for Natural Science and Technology (NCNST), which is now the Vietnam Academy of S&T, with functions to do not only fundamental but also applied oriented research.

However, this period lasted only a few years until the Asian crisis occurred in 1997, which severely affected Vietnam, especially in terms of FDI. Some initial achievements at the beginning of the 1990s were jeopardized.

More opening and integration: the road to innovation shift in a global context

At the end of the 1990s, the restructuring of the S&T system began with the elimination of subsidies to a number of governmental research organizations with Decision 782 in 1996. The reform process continued with the establishment of new institutions and organizations such as high-tech parks in Hanoi and Ho Chi Minh City. This period also began with more attention being paid to the promotion of innovation in the corporate sector via Decision 119 in 1999 on financial measures for innovation in small firms.

Further changes were institutionalized in the form of a number of new laws and policies, geared at promoting not only R&D activities in labs (Decision 850 to create 17 key national laboratories) but also more innovation activities in firms and enhancing linkages between firms and research organizations (Decision 214 on technology market development), or to make R&D organizations more autonomous in order to handle innovation (Decision 115). More importantly, many new laws have been put in place such as the first S&T law in 2000, intellectual property (IP) law in 2005, technology transfer law in 2006 and high-tech Law in 2008. For the long-term vision, the Strategy for S&T Development 2001–10 was approved in 2003 and the new S&T strategy for the next cycle of 2010–20 was passed in 2012.

After the year 2000, the economy has been consciously entering a new phase of development by integrating itself more actively into the international market system. The period is marked by the accession of Vietnam into the World Trade Organization (WTO) in 2006 and improved microeconomic management by introducing various laws and regulations. A range of new Free Trade Agreements are being negotiated or concluded with many countries (BTA with the United States is one example). The adoption in 2003 of the Science and Technology Development Strategy 2001–10 created a new phase in the transformation of Vietnam's S&T system and policy. For the decade of the 2000s, a shift from S&T focus towards more innovation became more visible.

The aforementioned laws and policies make the country more open and internationally linked to the global system of STI activities. The Vietnamese organizational and institutional structure has gradually evolved to become more innovation focused and more compatible with the global system. For example, the Law of Standard and Technical regulations in 2007 tried to put Vietnam closer to international systems of standards. The master plan of international integration again made the innovation approach more familiar to Vietnamese stakeholders. One of the most action-oriented measures for innovation is the creation of a number of programmes and funds, such as the National Foundation for S&T Development (NAFOSTED), and more recently, the National Technology Innovation Fund (NATIF) as well as the National Programme for Technology Innovation. To undertake these programmes of equal importance is enhancing the institutional capability of the government, particularly the Ministry of Science and Technology (MOST), the State management body for the formulation, coordination and implementation of S&T and innovation policy. Under MOST several bodies have been created such as the State Agency for Technology and Innovation (SATI), the Viet Nam Science and Technology Evaluation Centre (now the Vietnam Institute for S&T Evaluation, or VISTEC) and the National Agency for Technology Entrepreneurship and Development of Commercialization (NATECH). Next to MOST, there is the National Council for Science and Technology Policy (NCSTP), consulting body to the prime minister in providing direct advisory support to the highest level of leadership regarding STI. Increasingly, other organizations such as government ministries and research think tanks like CIEM also place more research focus on innovation.

3.3.3 Towards an Innovation System

As can be seen from the above, the STI system of Vietnam has been formulated over time, step by step, with various stages of institutional changes. However, shifting from a traditional S&T focus to a more innovation orientation took several decades.

The emergence of the innovation system concept and practice can be linked to a number of important international cooperation activities in this area. In 1982, the Swedish SIDA supported a delegation to Vietnam led by leaders of the Research Policy Institute (RPI) at Lund University (Jon Sigurdson and Claes Brundenius) to work with ISM and this led to long-term cooperation in S&T policy. This beginning of links to Nordic scholars in the context of international isolation gave Vietnam the first glimpse of a non-Soviet school of thought on the concept of STI policy and system research. The model and direction of ISM was very much reliant on lessons from RPI. This laid the foundation for further international cooperation that came later, geared more towards an innovation focus (Vu Cao Dam, 2008). During this time, via cooperation with international organizations such as UNESCO, International Institute for Applied Systems Analysis (IIASA), United Nations Economic and Social Commission for Asia and the Pacific (UN ESCAP), United Nations Development Programme (UNDP) and UNIDO, Vietnam became more familiar with the international concepts of the systemic approaches in S&T development and R&D management (Tran Ngoc Ca, 2011).

Another major move to innovation occurred with cooperation with the IDRC.[3] While Vietnam prepared its first draft of the Strategy for S&T 2001–10, Mr Pham Gia Khiem, Minister of Science, Technology and the Environment (later on becoming Deputy Prime Minister and Foreign Minister) asked IDRC to conduct an S&T review for Vietnam, similar to that undertaken for other countries using OECD and IDRC formats. The purpose of the review was to examine the Vietnam S&T status, following similar concepts and methodologies used in previous international studies. As such, it would form a more advanced international imprint with some reflection on other countries' experiences in a comparative perspective.

The international review was carried out by six members (Keith Bezanson, Geoff Oldham, Jan Annerstedt, Francisco Sagasti, Dennis Hopper and Kun Mo Chung) who spent three weeks in Vietnam in 1997 on the study. In addition, two other experts were invited to provide additional support. Jack Smith from the National Research Council of Canada provided a training course on methodology for strategy making, and Martin Fransman from Edinburgh University shared experiences of some East and Southeast Asian economies in STI policy.

To understand the Vietnamese status of the S&T system, the international review team interviewed and discussed with around 70 organizations, institutions, departments, firms and associations and roughly 320 Vietnamese S&T policy makers, policy implementers and people affected by the policies. The government assigned NISTPASS under the then Ministry of Science, Technology and Environment (now Ministry of Science and Technology) to be the Vietnamese counterpart for the study.

In addition to the general conclusions and findings on the status of the system, and its unquestionable impact on the knowledge and information input for Vietnam's policy-making process, the report specifically produced

16 specific recommendations on various aspects of the Vietnamese STI system (Bezanson et al., 1999). Despite some minor difficulties with dissemination of the report (mostly related to using the terminology 'innovation'), overall all recommendations have been turned into implemented actions and have had a great impact on changing the STI policy landscape in Vietnam. It is also useful to discuss the concept and terminology of 'innovation' itself in the Vietnamese context. It was with the IDRC mission that for the first time the notions of 'innovation' and 'innovation system' were officially put into the policy discourse of the Vietnamese policy-making scene. When the mission report was translated into Vietnamese, the term 'innovation' was translated as *doi-moi*, which is the same term used for the Renovation era after 1987. As such, to many minds innovation means Renovation again and is something that has been done in the past and there is nothing new about that. Over time, to avoid this repetition, a new term has been created for 'innovation' as *doi moi-sang tao* (literally meaning 'innovation-creativity'). By adding *sang-tao* (to create, or creativity), writers tend to emphasize more the novelty and research-related focus. But this has some drawbacks. To make it shorter, many in the media and writers began to use the term as *doi moi sang tao*, which to many could mean very different things (either 'innovate the creativity' or 'creatively innovate'). Ironically, when some Vietnamese writings were translated back to English, only the nuance of creativity survived and in some cases innovation just disappeared. This actually added more confusion to the discussion and led to difficulties in understanding the concept of the innovation system.

Despite these initial difficulties, increasingly innovation has become a familiar and accepted concept for the Vietnamese mindset, especially for policy makers and technocrat people. Gradually, S&T efforts moved away from the 'ivory tower' of only research institutes to serving more the needs of companies and society at large. As for universities, traditionally they were mainly fulfilling the teaching mandate, until becoming more research oriented recently (Tran Ngoc Ca, 2007). Most recommendations of the IDRC review report have been put into practice, such as the creation of a foundation mechanism for S&T funding, promoting peer review, using foresight in the planning processes, setting up national programmes to support small and medium-sized enterprises (SMEs) in technology innovation, to name a few.

Looking at what has been implemented, it appears that most of the recommendations in the IDRC report have contributed positively to the process of designing new policy measures for promoting STI in the country. These recommendations have found strong resonance from the community of Vietnamese policy makers (Tran Ngoc Ca, 2013). The practice of policy making for S&T in Vietnam has evolved along the path pinpointed in the review and reflects the focus of an innovation policy per se. Most importantly, there is a strong shift away from S&T policy to innovation policy with emphasis on enterprises, especially SMEs, which are central to any productive sector in Vietnam.

After this first review, another review on international cooperation in science and technology was also supported by IDRC. In 2002–04, the Department for International Relations of the Ministry of Science and Technology asked IDRC to assist their attempt to design a strategy for developing international cooperation and integration in science and technology (ICST). This was to support the formulation of new projects fostering international integration in S&T (later it was approved as the programme for science and technology international integration by the government). To conduct the project, a task force was set up and drafted the structure of the report, which looked at different components of the innovation system, such as human resources, financing, linkage with and support to enterprises, change of S&T management mechanism, link with FDI and ODA activities and so on.

During 2000–01, in a similar project supported by UNIDO/UNDP, Geoffrey Oldham and Keith Bezanson together with Tran Ngoc Ca prepared another study for a science, technology and industry strategy for Vietnam, as one component of an overall development strategy up to 2010–15 (Bezanson et al., 2001). This continued process of working together created unique opportunities for capacity building in Vietnam on STI policy making, and made innovation closer to the Vietnamese way of thinking, both in terms of content and methodology on how to study it.

At roughly the same time, some activities initiated by large multinational companies contributed to the process of awareness raising on innovation. In 2006, under the framework of the Asia-Pacific Economic Cooperation (APEC) Summit in Hanoi, IBM cooperated with MOST and the Vietnam Chamber of Commerce and Industry to organize the so-called Vietnam Innovation Summit. At this Summit, the concept of innovation and the national innovation agenda was raised, although through the lenses of a multinational company. The same efforts were seen in cooperation with a number of other large players like Intel, Nokia and Samsung.

All the above efforts have led to a stronger acceptance and embracing of the concept of an innovation system in Vietnam, resulting in more work being taken to a more advanced stage such as the Innovation Partnership Program in cooperation with Finland, most recently the FIRST project (Fostering Innovation through Research, Science and Technology) and

VIIP (Vietnam Inclusive Innovation Project) in collaboration with the World Bank.

3.3.4 Conclusion

The experiences of Vietnam reveal that S&T activities and the building up of the innovation system date back to the early 1960s, and have gradually evolved into an innovation system approach during the 1990s. Over time, the whole institutional setting of STI (including formulation of policies and creation of relevant organizations to implement these policies) came into play. The strong shift from an S&T notion to innovation aspects can be seen strongly today. In this context, collaborative activities with foreign partners have proved very useful for the country in the policy-making process. These cooperative efforts have provided background information, new knowledge on basic concepts, updating on latest trends and, most importantly, provided the analytical tools and framework that are strong catalysts for the innovation concept and practice to blossom. Without these efforts, there would be fewer changes in the overall institutional and organizational structure of Vietnam conducive for innovation.

However, the way these cooperative results (be they review, report or programme) have been responded to, studied, accepted and diffused may not be the same in every society and country. Depending on the policy opportunities (or window), the specific context and the cultural circumstances of each country, the recommendations of the collaborative efforts may be adopted and appreciated to different extents and in various forms. Experiences show that it takes time for a country like Vietnam to absorb and adopt the new vision suggested by external experts (innovation and the national system of innovation is just one example, among others) and turn them into own action by local efforts. Without this slow, but firm 'policy adoption and assimilation', attempts to make changes could be short-lived. One should take this into account and be ready to accept some kind of policy time lag.

There is no doubt that in a country like Vietnam, the emergence and development of an innovation system, which has taken place over the span of several decades, is slowly but gradually becoming mainstream thinking in policy-making circles.

3.4 THE EMERGENCE OF THE NATIONAL INNOVATION SYSTEM APPROACH IN TANZANIA

3.4.1 The Lagos Plan of Action 1980–2000: An Important Landmark for S&T Policy in Tanzania

The importance that has been given to S&T policies in Tanzania dates back several years to the early 1960s. In 1963 Tanganyika had participated actively in the United Nations Commission on Science and Technology for Development (UNCSTD) for the benefit of the less developed areas, held in Geneva, and in 1968 the Tanzania National Scientific Research Council was organized, although only inaugurated and operational in 1972. Later, under the auspices of UNESCO in 1974, the country participated in the First Regional Conference of Ministers Responsible for Application of Science and Technology in Africa (CASTAFRICA 1). In 1979 Tanzania again participated actively in the 1979 UNCSTD held in Vienna (AAPAM, 1993; Lim, 2011).

An important landmark in this early process occurred in 1980 with the adoption of the Lagos Plan of Action for Economic Development of Africa 1980–2000. The Plan in its S&T chapter is based on the African consideration of the negative results of the Vienna Conference, which in the view of Aguirre-Bastos and Gupta (2009) constituted an excellent opportunity for each country to examine their S&T situation.

Chapter V of the Lagos Plan of Action declares that the Vienna Conference realized only part of its objective and that the Vienna Programme of Action left much to be desired. In this context, the Plan calls for the African states to direct their efforts to 'spelling out a strategy for development which should guide their thinking, planning and action on bringing about socioeconomic changes necessary for improving the quality of life of the majority of the people'. Such an objective must be accompanied by 'investments in science and technology, calling at the same time the political will and courage on the part of policy and decision-makers of the continent to induce a profound change with far-reaching effects on the use of science and technology as the basis of socio-economic development as a matter of the utmost importance and urgency at this fateful juncture of history'.

The Lagos Plan of Action, after characterizing the situation of S&T in Africa, called for all member states to adopt measures to ensure its development in specific social and productive sectors of the economy, encompassing a large number of objectives and goals. The Plan recognizes that in spite of the fact that a number of institutions for S&T have been set up, member states lack the necessary national scientific and technological capability and consequently remain dependent on foreign technical skills for the execution of their scientific and technological tasks.

The Plan identifies a large set of factors that explain the shortcomings. From the policy perspective the Plan highlights the existing vagueness in the concept of S&T policy and its contents. In the discussion of the shortcomings, a first indication of an innovation system vision is brought forward, signalling that for S&T to play an effective role in national development, it must operate within a national framework and be coordinated with other sectors of the economy.

The Plan also calls for the establishment by each member state of a national 'Centre' for Science and Technology for Development. In this context the 'Centre' is not necessarily meant to connote a physical and monolithic structure. It is used as 'shorthand' for a national organ or mechanism, to define the way each member state can characterize the structure and modus operandi of its governmental machinery. The creation of such a mechanism was also pursued forcefully by UNESCO and several other multilateral and bilateral cooperation agencies operating in Africa in the 1980s and 1990s.

The Plan offers an insight into an innovation system by signalling that these Centres should have inputs and effective linkages with national public and private institutions, academic organizations and different levels of government as well as technological service organizations such as the industrial property system, technical and patent information, standardization and quality control systems. The Plan also envisages these Centres promote the integration of foreign technological inputs and those inputs that are generated by the indigenous S&T system and utilized by the national sectors of production and services. This is the first mention of a system in the Lagos Plan of Action.

The Plan calls for the Centres to develop capabilities for formulating an explicit national S&T policy that translate the national policy for socio-economic development into technological lines of action; it also calls for the creation of the missing components needed for a comprehensive S&T system (such as R&D institutions, technological information services, consultancy services, pilot plants and testing grounds, standardization and quality control establishments) and the identification of the needs of the national economy in terms of goods, services and techniques, which can be provided with the help of the national S&T system.

It is interesting to note that the Lagos Plan of Action calls for the establishment of policy instruments that encourage local innovations and facilitate more rapid and less costly acquisition of technology, through the provision of incentives and programming of the integrated and coordinated activities of the national S&T system.

3.4.2 The First National S&T Policy: 1985 and 1996

Under the influence of the Lagos Plan of Action and other external inputs, as well as the recognition by government of the importance of creating S&T capabilities, preparations for policy were initiated in 1984, which concluded in the definition of the first national S&T policy in 1985, later reviewed and reformulated in 1996 by the Ministry of Science, Technology and Higher Education.

To coordinate and implement the policy, the Tanzania Commission of S&T, COSTECH, was established in 1986 by Act of Parliament No. 7 as a successor to the Tanzania National Scientific Research Council. COSTECH became operational in 1988. The Act provides a structural framework of the Commission, which brings together the top leadership of the scientific and technological institutions in the country under one forum and is the principal advisory organ of the government on all matters relating to scientific research and technology development in the country, including the formulation of S&T policy, priority setting for R&D, allocation and utilization of resources (mandates modified by the creation of the then Ministry of Science, Technology and Higher Education in 1991).

The national S&T policy developed by the Commission and adopted by the government in 1996 was actually the second edition of the earlier policy. The policy set out a strategic direction for S&T. In hindsight, however, a component for innovation was largely missing, with S&T not seeming to be conceived as closely connected with and necessary for everyday economic life and changes in society.

The policy of 1996 was based on the recognition of several factors that were identified as limitations to the implementation of the 1985 policy. Of those more closely related to the national innovation system approach are the absence of linkages between R&D and local industries, even when several institutions such as universities, Tanzania Industrial Research Development Organization (TIRDO), International Press Institute (IPI), Tanzania Industrial Studies and Consulting Organization (TISCO) and Centre for Agricultural Mechanization and Rural Technology (CAMARTEC) had already been established.

Another factor that limited the development of S&T was identified as the poor structuring of R&D institutions, in particular their lack of vision towards streamlining their researchers' activities and to link them directly to current socio-economic issues. Also affecting the application of the 1985 policy was the poor funding and comprehension of the magnitude of the S&T phenomena. At the same time, it was recognized there was a lack of strategy to diffuse the use of technology and a lack of strategy to encourage and support the emergence of indigenous technologies.

A major issue at the time of revision of the policy was to consider whether the policy prescriptions were still valid, considering the changes that had taken place particularly since 1985 in the overall economic development strategy. Questions arose as to what extent the conceptual base of the policy is valid and what are its deficiencies. The limitation of the policy was identified as its inability to address the question of S&T in the context of the economic reform programmes, which were characterized by the objective of political and economic self-reliance. A coherent S&T policy was called for in the analysis.

The objectives of the 1996 S&T policy were to regulate the flow of technologies and reduce excessive dependence on imported technologies; to guide the development of national capabilities in S&T and define priority sectors and areas for R&D with the potential for accelerating development and competitiveness; to establish an effective institutional framework and linkages 'so that the various science and technology institutions and the productive sector work in a system that is focused on achieving developmental goals without undue duplication and competition'. As a result of the implementation of the 1996 policy, several policies and legal frameworks were put in place and new institutions, including new private universities, emerged.

In spite of the accomplishments, it is considered that the 1996 S&T policy fell far short of its objectives in its application. S&T could not impact the productive sectors nor result in commercial ventures. The central coordinating role sought for COSTECH was difficult to achieve, as R&D was, and still is, under each individual line ministry. One particular shortfall of the policy was the conception and subsequent practice of S&T that seemed to be an end in itself, without the important complement of innovation (as recognized in the policy review as discussed below).

3.4.3 Review of Policy in the 2000s

The subsequent key milestone in the S&T management in Tanzania was the publication of the S&T Sub-Master Plan (2003–18) in 2002, six years after the national S&T policy was adopted. The overall objective of the S&T Sub-Master Plan was to make operative the existing S&T policy by formulating clear issues and priorities, setting targets and timeframes, identifying resource requirements and establishing a system for monitoring and review. It was an S&T development plan expected to foster both the individual and collective development needs of the Tanzanian society. The 166-page S&T Sub-Master Plan was to be adopted as a government white paper before being implemented. However, as it turned out, the implementation of the Plan required a review of both the S&T policy and the legal framework.

In 2008 with the creation of the Ministry of Communications, Science and Technology, a 'Draft National Science, Technology and Innovation Policy' was prepared. This is the first formal document that introduces innovation as an element for future policy definition in the country.

The Draft recognizes that there is a close and mutually reinforcing relationship between science, technology and innovation; and where a combination of these exists and is upheld, STI can and will play a vital role in the self-sustaining development of any society. It is recognized that while Tanzania is rich in natural resources, like many other African countries, it is poor in the exploitation and utilization of science, technology and innovation in their proper combination; thus, the country is in need of a strong resolve to utilize scientific, technological and innovative capacities and capabilities to develop, transform and utilize the resources with which it is endowed.

Under such premises, and in view of the then formulated polices contained in the National Development Vision 2025, the United Nations Millennium Development Goals (MDGs), the National Strategy for Growth and Reduction of Poverty (NSGRP/MKUKUTA) and the Chama Cha Mapinduzi (CCM) Party Manifesto, a new policy was put forward in the Draft. The policy 'Vision' was set 'to acquire the national capacity and capability to harness and apply science, technology and innovation for sustaining livelihood and wealth creation'. Key to the discussion in this chapter, the mission was set 'To create an efficient, effective and conducive scientific and technological milieu and innovation system that is responsive to the needs and aspirations of Tanzanians and ensure a self-sustaining, competitive and growing economy.'

Five policy objectives were defined to guide the application of the proposed national policy:

a. Giving STI the highest possible national prominence.
b. Mainstreaming STI into all sectors of the national economy.
c. Promoting a scientific culture across the board.
d. Creating a mechanism for setting up and periodically reviewing the national STI agenda.
e. Creating participatory mechanisms for reconnaissance, foresight, planning, implementation, monitoring and evaluation system in STI-related issues.

Several strategic objectives were defined to accomplish the main objective, some related to the institutional governance framework, including: create a separate ministry for STI, empowering COSTECH to finance STI activities in tandem with the STI agenda; set up an STI system based on clear and objective criteria to ensure that STI management receives informed and well-articulated ideas from the respective fields;[4] ensure public funding for STI; establish an enabling environment for intellectual property rights (IPR), technology incubation, clusters and licensing systems for effective technology innovations, development, transfer, indigenous efforts and/or commercialization.

The Draft calls to facilitate and support the establishment of a national innovation system, creating indigenous knowledge systems as well as using new and emerging technologies in the areas of ICT, biotechnology, new materials, nanotechnology and nuclear science.

In May 2010 the Ministry of Communications, Science and Technology (MCST) adopted a new Master Plan, and within it a 'Science, Technology and Innovation' component.

The Master Plan 2010–20 recognizes the country's weaknesses and limitations in STI and aims at setting out a long-term plan that will guide the implementation of the mandate of MCST in the next decade. The main aims of the Plan are:

a. To realign MCST with current and future national priorities, immediate needs of society, ongoing public sector reforms, and stakeholders and market demands.
b. To accelerate the development of STI and harness its full potential and that of ICT to improve education and innovation, public sector management, private sector competitiveness in key strategic productive areas, industrialization, capacity building and poverty reduction in line with key sustainable development objectives as set in MKUKUTA, National Development Vision 2025, the MDGs, as well as other national development policies and strategies.

The Plan defines the MCST long-term goal in the next decade as a catalyst that will enable the country to raise its productivity and competitiveness, and to improve the standard of living and quality of life of Tanzanians through promotion and effective use of S&T, information and innovation processes, products and services in both productive and non-productive strategic sectors of the economy.

It defines that STI will be used in strategic areas of the economy to enable the acceleration of real GDP growth from the current 5 per cent to 9 per cent by 2020. MCST areas of strategic focus in 2010–20 are: food security; high value, high demand processed agricultural and natural resources, products and services; education and STI capacity and capabilities development; water; energy; infrastructure-technical and informational; and health.

Besides these seven strategic areas the Plan also defines other priorities to be pursued: indigenous knowledge and technologies; environment management; industrial development; ICT; social sciences; basic research in the fields of agro-biotechnology, medicinal biotechnology, profitable and high-utility emerging STI and its applications, for example, nanotechnology; R&D in trade.

It is interesting to note that the Master Plan describes its Vision thus: '[T]he sustainable economic growth and social development of Tanzania are driven by science, technology and innovation'. Its mission is defined as: 'To be the leader in mobilizing and empowering the economic agents to embrace and utilize science and technology through improving their knowledge base, technical skills and expertise to enhance sustainable development and quality of life.' The objective of setting up the national innovation system as expressed in the earlier Draft is no longer mentioned in this Master Plan.

3.4.4 The National Research and Development Policy (June 2010)

Derived from the Master Plan, the new policy recognizes that research plays a crucial role in social and economic development, and that Tanzania has been undertaking significant research efforts in some areas, but results have not been translated into tangible products, processes and services for development purposes. The policy was actually adopted to be part of and complement the 1996 policy, which was based on a supply side linear model with little emphasis on innovation and the role of market forces, while at the same time it is embedded in the present social and economic policies and the National Development Vision 2025, with the final aim to develop a competitive knowledge economy, which is skill-based, knowledge and innovation driven, and thus capable of generating and sustaining dynamic development.

A large number of factors that hinder the utilization of research results are recognized by this policy document, in particular the weak multidisciplinary interaction and collaboration between R&D organizations, weak social and economic research, inadequate mechanisms for including new and emerging research areas into the national research agenda and low participation of the private sector in research activities.

These limitations are exacerbated by inadequate human resources, overdependence on foreign funding and inadequate appreciation of the role of research in national development. In such a context the policy is intended to strengthen the research capacity in order to bring about increased competitiveness and provides guidelines on how research should be conducted in all sectors of the economy in order to:

a. Establish an appropriate coordination and management system of research activities.
b. Set clear and realistic priorities for research on a short, medium and long-term basis.
c. Rationalize the use of the country's natural resources through R&D activities.
d. Request that government increases its budgetary allocation to research and sets guidelines for increasing the share of the private sector in funding R&D.

It is interesting to note that in application of these objectives, the health sector innovation system was actually constituted by an expensive collaboration between COSTECH and the health sector institutions (de Haan et al., 2015). Further, different researchers have used the innovation system approach to analyse and recommend policy options. There are several examples including in-depth analysis of the agriculture sector innovation system (Diyamett et al., 2011; Mwamila and Diyamett, 2009; Paterson et al., 2003).

3.4.5 Conclusions

It is recognized that there exists a set of key policy issues in Tanzania that need to be addressed (see, for example, Lim, 2011 and the extensive work of the Science, Technology and Innovation Policy Research Organization, STIPRO). The utilization of the innovation system approach has certainly great potential to address challenges posed by such issues.

In general, since the early 2000s, the national innovation system approach has become embedded in the government's discourse. For example, in July 2007, the president of the United Republic of Tanzania requested UNESCO's director general to assist Tanzania to reform and reposition the country's STI system in a way that would enhance the contribution of the sector to national economic development. But this 'purpose' is not reflected in the 2010 national STI policy.

Besides recognizing the weaknesses and strengths of STI in Tanzania, one of the important findings of the study of Aguirre-Bastos and Rebois (2011) in their analysis of the STI system is that the MCST sets policy and oversees STI organizations that are more powerful than itself (which is recognized as the assigned role of COSTECH). Further, sector R&D policies are defined by sector ministries, and each sector administers their

R&D organizations, so the question is then, what does the MCST actually administer?

It is also recognized in the above study that one key issue is that of the STI system or national innovation system approach itself. Mention of the system is done in several policy documents, some of which actually define the establishment of the system as objectives. But this purpose is readily changed in the next policy document. In other words, although the national innovation system approach is called for in several opportunities, no consistent use of its concepts is evident from the different government policy documents.

In March 2015, the government called for a three-week living laboratory workshop to review the national STI system in the country. The conclusions of the workshop were to become available later in 2015. It is to be expected that the conclusions will lead to the full understanding and strengthening of the innovation system approach and its policy adoption and implementation in the country.

3.5 THE EMERGENCE OF THE INNOVATION SYSTEM APPROACH IN MOZAMBIQUE

3.5.1 The Context

Mozambique became independent in 1975, but only in 1992 was the country able to reach a peaceful political agreement that provided the basis for economic and social development. Since 1992 the country has experienced stability and a high rate of economic growth. Today it is not only a stable democracy but one of the fastest growing economies in Africa (7 per cent in 2005–2013[5]).

In spite of this growth, Mozambique remains a poor country heavily dependent on foreign aid. The impact of the dependence of Mozambique on foreign aid has been amply documented. It is generally considered that there is a diminishing trend in impact due to improvements in State revenues, especially with the presence of megaprojects in the natural resources sector and due to the improvement of the tax collection system that has contributed to reduce budget deficits.

The evaluation of budget support operations undertaken from 2005 to 2012, made by ITAD, FISCUS and MBConsulting (2014), commissioned by the Evaluation Unit of the European Commission, states that budget support has effectively strengthened the structure for policy dialogue, based upon the definition of policy targets. The evaluation recognizes at the same time that the main structural weakness of the policy dialogue process lies within the policy development process itself. Although the evaluation does not deal with S&T directly, it does refer to closely related issues such as the need to increase productivity levels in agriculture and to improve the environment for doing business. It is clear that the above observations require the notice of S&T policy makers.

In relation to foreign aid, Torvinen (2013) discusses the issue of governance, concluding that the country is still unable to take its own decisions due to the external influences it suffers from donors. This and other assessments make very little reference to issues related to the governance of science, research and innovation.

It is certainly one of the country's great challenges how to put its economic growth at the service of poverty reduction and social inclusion, especially in the rural areas. Different national development policies and plans call on S&T to deliver responses to this challenge (see, for example, Massarani and Lima, 2012).

3.5.2 Institutional Developments in Science, Technology and Innovation

As is the case for most African countries, S&T in Mozambique is no stranger to its development plans. In 1962 during the colonial period the Centre for University Studies (although almost strictly for training Portuguese citizens living in Mozambique) was established, becoming the Eduardo Mondlane University after independence and being the only higher education institution between 1975 and 1993. The functions of the university in the area of S&T are described by UNESCO (1990) as:

a. advocacy of S&T (highest priority)
b. promotion, financing and coordination of scientific and technological activities; application, transfer, assessment of S&T (high priority)
c. general policy advice (low priority).

The university was linked upstream to the National Council of Education and the Council of Ministers and downstream to its faculties and research centres. It acted as the Secretariat of the then existing Scientific Council. The UNESCO Directory states the university had rare professional contacts with institutions in foreign countries dealing with S&T studies.

In 1999 the Ministry of Higher Education, Science and Technology (MHEST) was established as a result of a national consultation that the country's president requested on the need to expand higher education. This was the first time a specific ministry in charge of S&T had existed in Mozambique and provided an important first formal step taken by government to promote and develop S&T. In 2003 the Ministry defined a policy proposal that described further below.

In 2005 Presidential Decree No. 13/2005 of 4 February created the Ministry of Science and Technology (MST) and Presidential Decree No. 17/2005 of 31 March established its objectives and functions. The latter were determined as:

a. formulation of policies and strategies for the development of S&T
b. define norms, standards, regulations and coordination in the S&T areas
c. planning, monitoring and analysis of the development of S&T
d. promotion of scientific research and technology innovation
e. promotion of the diffusion of science
f. promotion of valuing local knowledge and its diffusion
g. promotion of protection of IPR
h. promotion of research and innovation methodologies based on professional ethical values and that secure benefits for the economic, social and cultural development of the country
i. promotion of development through the introduction of new and disruptive technologies
j. coordination of the activities of research and technology development.

In the period 2005–13 an important set of regulatory tools, including the creation of various coordination institutions, funding agencies and different sectorial policies were established, and are today pillars of the development of STI making up the national system of innovation as discussed further below.

Among the policies, strategies and regulations are the S&T policy; the ICT policy; the Mozambique Strategy on STI; the Strategy of Intellectual Property; a Strategic Plan for Higher Education; Plan of Human Resources Development; regulation on biodiversity; regulation on the researchers' carriers; Code of Ethics; National Mobility System of Researchers; Regulation for Licensing the R&D Activity and Research Institutions Register.

In this period important coordinating institutions were established: the National Council for Science and Technology (CNCT) – the highest coordination organ, created by the Council of Ministers, whose chair is the minister of S&T; the Mozambique Academy of Science; and three thematic Scientific Councils for Water, Energy, Health, Agriculture and Ethnobotanics. Funds to promote S&T were also defined and put into operation, mainly: the National Research Fund; Agriculture Research

Competitive Grants Project; Agriculture Development Fund; Funds for Quality and Innovation; Fisheries Research Fund: Open Fund from Eduardo Mondlane University (UEM).

Also, new public research institutions were either strengthened or created, as well as higher education institutions. MST also has representation in regional centres. Finally, on 16 January 2015 the Ministry of Science and Technology, Higher and Technical Education was created by Presidential Decree No. 1/2015. The Ministry is reorganizing itself to play its new roles.

3.5.3 S&T Policy Developments

Because of its colonial dependence before 1975 and later the civil strife in the country between 1975 and 1992, Mozambique – in contrast to many other countries in the region – could not be a very active player in the regional and international forum on S&T, but was influenced by their outcomes. An important example of the latter were the two regional ministerial conferences organized by UNESCO: CASTAFRICA I (Dakar, Senegal, 1974) which identified and recommended, among other issues, the need for African countries to formulate their S&T policy to guide national development, especially rural and industrial development and CASTAFRICA II (1987) that took place in Arusha, United Republic of Tanzania in 1987, which produced a large number of recommendations on the ways the African nations could take in order to manage S&T. The country also drew from other key regional consultations on S&T, such as the Lagos Plan of Action in 1980, which in general influenced the definition of S&T policies in Africa (as discussed further in Chapter 6).

It is only after 1992 that the country started to benefit from the interactions with policy-making institutions and experts in S&T. As an example, a source of inspiration for policy development and later the innovation system approach has been the extensive work of the African Technology Policy Studies Network (ATPS). The bilateral cooperation provided by SIDA (Sweden) and FINNIDA (Finland) were also pivotal in introducing S&T issues in the government, academic and productive structures of the country (see, for example, Elming and Abrahamsson, 2010). In this process the individual contribution of national researchers also has to be highlighted.

In 2003 after a set of seminars, consultations at national and regional levels, studies, round tables involving a large set of actors, MHEST defined a national plan for S&T (MHEST, 2003). The policy has had a noticeable continuity, as it is considered that it provided the basis for the

Mozambique Science, Technology and Innovation Strategy (MOSTIS) in 2006 that is currently being applied in the country.

The stated mission of the 2003 policy was set as: 'The policy should contribute to the reduction of absolute poverty, to improve the standard of living of the population and to national development, through the creation of conditions to foster research, encourage and promote scientific and technological competences, including local knowledge, and through the promotion of more effective processes of technology transfer and endogenous technology production.'

The analysis that provided the main inputs to the 2003 policy was resumed in an extensive SWOT analysis. Some of the features of the SWOT that are relevant to this discussion are:

a. Strengths: the existence of the MST as a coordinating and promoting government agency; a set of research and higher education institutions, both private and public, in expansion or being reformed; growing number of enterprises with R&D capacity; national and international researchers exchange.
b. Weaknesses: scarce scientific production and lack of research outputs for policy decisions; lack of linkages between social media and S&T actors and other production agents; weak popularization mechanisms; local communities with ancestral knowledge are not part of the R&D establishment. Weak infrastructure and lack of financial resources for research, making this activity dependent on outside sources, are two critical problems identified in the SWOT analysis. Limited research activities in the enterprise sector are not conducive to technology development.
c. Opportunities: the new climate of peace and the economic growth; rich natural resource base; diversity of local knowledges; the execution of mega projects, which have been the key to the high growth observed throughout the years.
d. Threats: lack of a legal framework and incentives that promote competence and quality; relevance of research institutions; lack of financial resources.

Of relevance to this chapter, the SWOT analysis recognized that the national research and development system was only at a very early stage of definition; it also recognized the outdated legal framework and financial compromise to allow research institutions to operate successfully. This is the first reference to the system approach, and shows that policy makers were aware of it and understood its basic operational principles and the need to adopt it.

In fact, the stated main objective of the policy was to develop an integrated system of production and management of knowledge addressing national needs as a way to promote the sustainable development of the country. More specifically, the policy called to:

a. develop an integrated, dynamic and quality research and innovation system
b. strengthen the national education system and professional training in the component of the creation of S&T capacities
c. develop the innovative capacity of the national productive sector
d. develop a dissemination and communication system of S&T knowledge, taking advantage of the new ICTs.

The policy contains a large set of strategic lines. Those contained within the framework of the suggested integrated system are:

a. strengthen the linkage – between the system and civil society
b. promote linkages between research activities and the productive sector
c. create the National Academy of Sciences
d. stimulate the expansion of research towards other areas of knowledge of importance to national development
e. stimulate the expansion of research institutions throughout the national territory
f. stimulate the participation of women and the young in research
g. create and implement an indicator system for measuring the performance of research
h. stimulate the research and use of local knowledge
i. stimulate excellence at the individual, collective and institutional levels
j. create technology reference centres.

On the basis of the 2003 policy, MOSTIS was later approved in June 2006. Elming and Abrahamsson (2010) consider that the Strategy is a 'reflection of a good ambition to develop research and built research capacity in Mozambique'. One key point raised in the Strategy is the importance of the process by which individuals and groups 'devise new ways to solve immediate problems' (Ministry of Science and Technology, 2006, p. 55).

MOSTIS contains seven strategic objectives (Nhampossa, 2013):

a. foster a culture of innovation throughout Mozambican society
b. promote STI within the industrial and public sectors and promote the transfer of technology

c. promote the use of ICT for good governance and service delivery, and for the diffusion of knowledge, in support of poverty reduction and economic growth
d. promote human resource development at all levels in the areas of science, technology and innovation
e. build and improve the policy instruments, institutions and infrastructure of the S&T system
f. establish funding policies and mechanisms for research and innovation
g. promote the mainstreaming of S&T within all sectors.

Nhampossa (2013) states that MOSTIS establishes the national system of S&T and is to be consolidated through research and innovation programmes but maintains the autonomy of individual institutions.

Walters-Monteiro (2010) is very critical of the implementation of the policy efforts in Mozambique, signalling that 'despite the emphasis given, in terms of policies, to the importance of science and technology to the development, the practical actions show that the existing habitus in the education system, conditioned by a colonial legacy where the knowledge is transmitted in a limited form and, by the absence of a scientific culture that affects the S&T system and some segments of Mozambican society, contributes to the maintenance of a vision of small relevance of the scientific and technological activity to the development of the country'. Such a critical view can actually be extrapolated to the large majority of situations in developing countries where the rhetoric of the importance of STI is not matched by real commitments to its development.

3.5.4 The Making of the Innovation System Approach

As is the case for most developing countries, the innovation system approach in Mozambique emerged from a series of occurrences, some of which have been mentioned above. In the particular case of UNESCO, its activities in Africa have been to develop new approaches for the formulation of science policies, including the innovation system approach (for an overview of the organization's contribution, see Bokova, 2012). Particularly important today is the development of the UNESCO Global Observatory on Science, Technology and Innovation Policy Instruments (GOSPIN) as it provides valuable information on national experiences in the formulation of policies. Several African countries have already developed the instrument with much success.

On the other hand, several researchers have emphasized the approach for the analysis and evaluation of the situation of research and innovation in the country. An interesting and relevant case is that of Davis et al.

(2008) that examines how post-secondary agricultural education and training (AET) can contribute to agricultural development by strengthening the capacity to innovate – 'to introduce new products and processes that are socially or economically relevant to smallholder farmers and other agents'. Using the AET system in Mozambique as a case study, the authors examine it within the context of an agricultural innovation system. This innovation system perspective, as called forth by Davis et al., offers an analytical framework to examine technological change in agriculture as a complex process of interactions among diverse actors who generate, exchange and use knowledge, conditioned by complex social and economic institutions.

The innovation system approach is also addressed in the study of Baltazar and Lopes (2011). The study discusses user innovation in firms from indicators collected in the National Innovation Survey of 2009, using the model questionnaire of the survey to gather official statistics in the European Union Community (Community Innovation System) that follow the guidelines of the Oslo Manual as adjusted by the Bogota Manual for the case of developing countries.

While the observations are too few to support statistical inferences, the authors demonstrate the presence of user innovation in firms and they suggest further work on the policy implications, especially related to support for innovation in firms that perform no R&D. The case study found little evidence of the sharing of knowledge gained through user innovation, in contrast with findings from industrialized countries, but closer to results from the newly industrialized countries such as Korea.

One of the first formal recognitions of the existence of the STI system comes from the publication on indicators produced by MHEST (2005). This publication 'is structured according to a specific vision about the process of scientific and technological innovation'. In such a vision, innovation is considered as produced by a system, composed of four sub-systems as shown in Figure 3.2.

As shown in the figure, the sub-systems are, respectively, research, education, production and dissemination. It is quite interesting to contrast this vision with that of the national system of S&T development constituted in Bolivia in the late 1970s as it contains almost the same four sub-systems.

The present system as understood today is described in Chapter 7.

3.5.5 Conclusions

The emergence of the innovation system approach in Mozambique came some time later than in many developing countries. It was triggered by extensive discussions at the national and regional levels and the

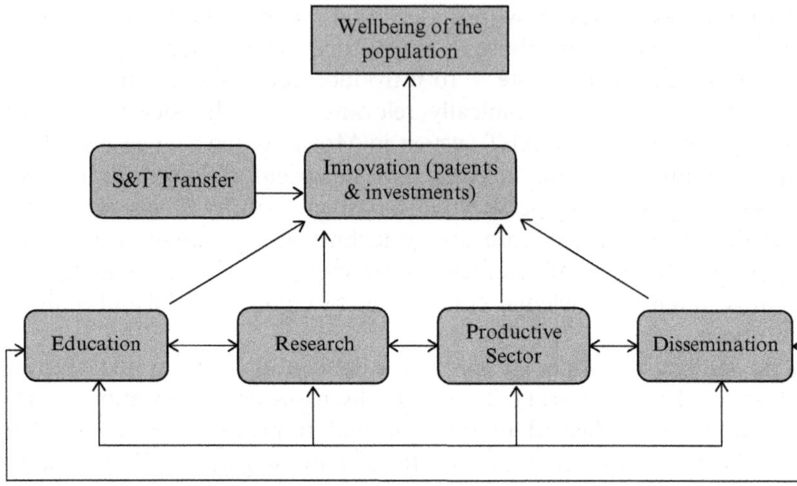

Source: Adapted from MHEST (2005).

Figure 3.2 The science and technology innovation system in Mozambique

contribution of several national researchers' work. As such, and different from other countries in Africa, it has a very solid potential base for its build-up.

The innovation system as defined in Mozambique has introduced definitions and concepts from the more developed countries and added some local ingredients, such as the case of the local communities as strong actors. It remains to be seen whether this approach to establishing the system will permit its operability.

Despite the government's efforts in developing various initiatives for S&T policy implementation, in creating legal tools and research institutions, and mobilizing financial resources, it is recognized that the Mozambique Innovation System (MIS) is still new (Ministry of Science and Technology, 2014) and much effort needs to be made in order for it to operate successfully.

NOTES

* The sections are written by the authors as follows: Section 3.1 is by Claes Brundenius, Section 3.2 is by Carlos Aguirre-Bastos, Section 3.3 is by Tran Ngoc Ca, Section 3.4 is by Carlos Aguirre-Bastos and Bitrina Diyamett and Section 3.5 is by Carlos Aguirre-Bastos and Maximiano Dgedge.

1. In contrast to the Sabato triangle in which the impetus to innovation is expected to come from government, the triple helix model posits multiple sources of initiatives arising from each sphere individually and in collaboration with one or two others.
2. The CINDA model emerged from the study, 'The System of Scientific and Technological Development in the Andean Sub Region' developed by Joaquín Cordua (Chile), Eduardo Aldana (Colombia), Máximo Vega (Peru), Álvaro de la Barra (Chile), Jaime Lavados M. (Chile), Hernán Ayarza (Chile) and Guillermo Ramírez (Chile). The proposal was first considered in an expert meeting held in Lima, Peru, at the headquarters of the Andean Pact Secretariat in April 1976, in which participated besides the above authors Carlos Aguirre-Bastos and Miguel Tejada Velasco from Bolivia. This proposal and subsequent model are considered here as a precursor of the national innovation system approach later explicitly developed, among others, by Lundvall and Edquist in the 1980s.
3. The text used Tran Ngoc Ca (2013).
4. This is referring to a deliberate setting up of a system involving 'think tanks' made up of respected, well-accomplished scientists and technologists in the country for the purposes of making effective use of them in situations that require critical counsel and inputs of 'the best minds', and where such individuals, in their own professional career, have made significant contributions in related research issues and/or product development processes.
5. World Bank database (www.worldbank.org/statistics).

REFERENCES

AAPAM (1993), 'Tanzania Country Paper: Status Review of Science and Technology Management', African Association for Public Administration and Management (AAPAM), 14th Round Table Conference.

Aguirre-Bastos, C. (2001), 'The Latin American policies', in P. Roffe and A. Yusuf (eds), *International Technology Transfer: The Origins and Aftermath of the United Nations Negotiations on a Draft Code of Conduct*, London: Kluwer Law International, pp. 229–36.

Aguirre-Bastos, C. (2002), 'El desarrollo del marco institucional y de políticas de la ciencia y la tecnología en Bolivia', in C. Aguirre-Bastos (ed.), *Contribuciones a la Historia de la Ciencia en Bolivia. Academia Nacional de Ciencias de Bolivia e Instituto Panamericano de Geografía e Historia*, La Paz, México: National Academy of Sciences of Bolivia and Pan American Institute for Geography and History, pp. 3–12.

Aguirre-Bastos, C. and M. Gupta (2009), 'Thirty years of Vienna and ten of Budapest: where are we in Latin America?', Editorial note, *Interciencia*, **34**(8), August, 865–72.

Aguirre-Bastos, C. and R.R. Rebois (2011), *Review and Evaluation of the Performance of Tanzania's Higher Education Institutions in Science, Technology and Innovation*, Final report to UNESCO, Dar es Salaam, 23 June.

Baltazar Zita, J.E. and H.A. Lopes (2011), 'User innovation in the business enterprise sector of Maputo province in Mozambique', UNU-MERIT Project on Research and Training Support to Build African Capacity in Science, Technology, and Innovation Indicators, IDRC.

Bezanson, K., J. Annerstedt, K. Chung, D. Hopper, G. Oldham and F. Sagasti (1999), *Vietnam at the Crossroads: The Role of Science and Technology*, Ottawa: IDRC.

Bezanson, K., G. Oldham and Tran Ngoc Ca (2001), Vietnam science, technology and industry strategy', Prepared as part of the UNDP/UNIDO project DP/

VIE/99/002, 'Vietnam – Contribution to the Preparation of the Socio-economic Development Strategy to the Year 2010', National Political Publishing House, Hanoi.

Bokova, I. (2012), 'An integrated policy approach in science, technology and innovation for sustainable development: a UNESCO idea in action', in S. Dutta (ed.), *The Global Innovation Index 2012*, Fontainebleau: INSEAD and Cornell University, pp. 143–8 .

Brundenius, C., B.-Å. Lundvall and J. Sutz (2009), 'The role of universities in innovation systems in developing countries', in B.-Å. Lundvall, K.J. Joseph, C. Chaminade and J. Vang (eds), *Handbook of Innovation Systems and Developing Countries*, Cheltenham, UK and Northampton, MA, USA: Edward Elgar, pp. 199–228.

CASTAFRICA II (1987), *Final Report of the Second Conference of Ministers Responsible for the Application of Science and Technology to Development in Africa*, Arusha, United Republic of Tanzania.

CINDA (1977), 'El sistema de desarrollo científico y tecnológico de la SubregiónAndina', Centro Interuniversitario de Desarrollo Andino (now Centro Interuniversitario de Desarrollo), Santiago de Chile, available at http://www.cinda.org (accessed 15 September 2015).

Davis, K.E., J. Ekboir and D.J. Spielman (2008), 'Strengthening agricultural education and training in sub-Saharan Africa from an innovation systems perspective: a case study of Mozambique', *Journal of Agricultural Education and Extension*, **14**(1), 35–51.

de Haan, S., R. Kingamkono, N. Tindamanyire et al. (2015), 'Setting research priorities across science, technology and health sectors: the Tanzania experience', *Health Research Policy and Systems*, **13**, 14.

Diyamett, B.D., A. Wangwe, A. Komba and E. Mneney (2011), *Science, Technology and Innovation for Poverty Reduction in Tanzania: An Analysis of the Sectoral Systems of Innovation*, Part I, Final report to ATPS, Tanzania.

Eklund, M. (2007), *Adoption of the Innovation System Concept in Sweden*, Uppsala Studies in Economic History, Vol. 81, Uppsala: Uppsala University.

Elming, S.-A. and K. Abrahamsson (2010), 'The National Research Fund (FNI) as a tool for research capacity building in Mozambique: an in-depth assessment of SIDA support to the Ministry of Science and Technology MCT, within the Research Cooperation with Mozambique', *SIDA Review*, 15.

Gault, F. (2013), *Handbook of Innovation Indicators and Measurement*, Cheltenham, UK and Northampton, MA, USA: Edward Elgar.

Godin, B. (2010), 'National innovation system: a note on the origin of the concept', Project on the Intellectual History of Innovation.

Göransson, B. and C. Brundenius (eds) (2011), *Universities in Transition – The Changing Role and Challenges for Academic Institutions*, New York: Springer.

Halty, M. (1979), *Technological Development Strategies for Developing Countries*, Montreal: Institute for Research on Public Policy.

ITAD, FISCUS and MBConsulting (2014), *Independent Evaluation of Budget Support in Mozambique: Final Report*, Vol. I, European Commission.

Jaramillo, L.J. and C. Aguirre-Bastos (1989), *La Otra Integración: Ciencia y Tecnología en la Construcción del Grupo Andino, evolución, alcances y perspectivas recientes*, Lima: Marens Editores.

Lim, M. (2011), 'Science, Technology and Innovation (STI) agriculture and

development relevance for Tanzania and Africa', UNCTAD, VI Study Tour, Geneva, 21 June.
Lundvall, B.-Å. (1985), *Product Innovation and User–Producer Interaction*, Aalborg: Aalborg University Press.
Lundvall, B.-Å. (1992), *National Systems of Innovation. Towards a Theory of Innovation and Interactive Learning*, London: Pinter.
Lundvall, B.-Å., K.J. Joseph, C. Chaminade and J. Vang (eds) (2009a), *Handbook of Innovation Systems and Developing Countries*, Cheltenham, UK and Northampton, MA, USA: Edward Elgar.
Lundvall, B.-Å., J. Vang, K.J. Joseph and C. Chaminade (2009b), 'Innovation system research and developing countries', in B.-Å. Lundvall, K.J. Joseph, C. Chaminade and J. Vang (eds), *Handbook of Innovation Systems and Developing Countries*, Cheltenham, UK and Northampton, MA, USA: Edward Elgar, pp. 5–24.
Martinez Vidal, C. and M. Mari (2002), 'La Escuela Latinoamericana de Pensamiento en Ciencia, Tecnología y Desarrollo: Notas de un proyecto de investigación', *Revista de la Organización de los Estados Iberoamericanos*, No. 4, September–December.
Massarani, L. and L. Lima (2012), 'Brazil–Mozambique links in scientific policy: an interview with Lidia Brito' (former Minister of Science and Technology of Mozambique), *Historia, Ciencias, Saude – Manginhos, Rio de Janeiro*, **19**(2), April–June, 533–40.
MHEST (2003), *Política de Ciencia e Tecnología*, Maputo: Ministry of Higher Education, Science and Technology, June.
MHEST (2005), *Indicadores de Ciencia e Tecnologia em Mozambique*, Maputo: Ministry of Higher Education, Science and Technology.
Ministry of Science and Technology (2006), *Mozambique Science, Technology and Innovation Strategy (MOSTIS)*, Maputo: Ministry of Science and Technology.
Ministry of Science and Technology (2014), *Mozambique Innovation System (MIS)*, Draft, Maputo: Ministry of Science and Technology, June.
Mwamila, L.M. and B. Diyamett (2009), 'Universities and socioeconomic development in Tanzania: public perceptions and realities on the ground', *Science and Public Policy*, **36**(2), 85–90.
NEPAD (2011), *African Innovation Outlook 2010 I*, Johannesburg: NEPAD.
NEPAD (2014), *African Innovation Outlook II*, Johannesburg: NEPAD.
Nelson, R. (ed.) (1993), *National Innovation Systems: A Comparative Analysis*, New York: Oxford University Press.
Nelson, R. and S. Winter (1982), *An Evolutionary Theory of Economic Change*, Cambridge, MA: Harvard University Press.
Nhampossa, E. (2013), 'Science, technology and innovation in Mozambique', Paper presented at the Kenya Institute of Education, Nairobi, 29–30 January.
OECD (1997), *National Innovation Systems*, Paris: OECD.
OECD (2015), Frascati Manual 2015, *Guidelines for Collecting and Reporting on Research and Experimental Development*, Paris: OECD.
Paterson, A., R. Adam and J. Mullin (2003), 'The relevance of the national systems of innovation approach to mainstreaming S&T for development in NEPAD and the AU', Draft Working Paper for the Preparatory Meeting of the First NEPAD Conference of Ministers and Presidential Advisor for S&T, Nairobi.
Prebisch, R. (1949), 'El desarrollo económico de América Latina y algunos de sus principales problemas', mimeo, CEPAL, Santiago, Chile. (English version:

'The economic development of Latin America and its principal problems', Lake Success, 1950.)
Rosenberg, N. (1982), *Inside the Black Box: Technology and Economics*, Cambridge: Cambridge University Press.
Sabato, J. (1964), Plan de actividades del Departamento de Metalurgia de la Comisión Nacional de Energía Atómica, Argentina', Paper presented at Conferencia de Expertos Latinoamericanos en Metalurgia de Transformación, CNEA, OEA, INTI and US Air Force Office of Scientific Research, Ciudad de Buenos Aires, 18–23 August.
Sabato, J.A. (ed.) (1975), *El pensamiento latinoamericano en la problemática ciencia-tecnología-desarrollo-dependencia*, Serie Mayor Política y Sociedad, Buenos Aires: Ed. Piados.
Sabato, J. (1994), 'El origen de algunas de mis ideas', in H. Ciapuscio (ed.), *Repensando la política tecnológica. Homenaje a Jorge A. Sábato*, Buenos Aires: Nueva Visión, pp. 11–74.
Sabato, J.A. and N. Botana (1968), 'La ciencia y la tecnología en el desarrollo futuro de América Latina. Estudio prospectivo sobre América Latina y el orden mundial en la década de 1990', Paper presented at the World Order Models Conference, Bellagio, Italy, 20–25 November.
Sagasti, F. (1983), 'La política científica y tecnológica en América Latina, un estudio del enfoque de sistemas', *Jornadas 101*, El Colegio de México, México DF.
Sagasti, F. and J. Hodara (1981), 'Maximo Halty y el pensamiento latinoamericano sobre política científica y tecnológica', *Comercio Exterior*, Mexico, **31**(5), May, 564–7.
Schumpeter, J. (1943), *Capitalism, Socialism and Democracy*, Floyd, VA: Impact Books.
Seidel, R.N. (1974), 'Toward an Andean Common Market for Science and Technology, Program on Policies for Science and Technology in Developing Nations', Cornell University, Ithaca, NY.
Solow, R. (1957), 'Technical change and the aggregate production function', *Review of Economics and Statistics*, **39**(3), 312–20.
STISA (2014), *Science, Technology and Innovation Strategy for Africa 2024*, Addis Ababa: African Union Commission.
Torvinen, L. (2013), 'Assessing governance assessments: the case of Mozambique', Doctoral Dissertation, Maastricht University, 21 November.
Tran Ngoc Ca (1999), 'Reforming the science, technology, education and training system in a transitional economy: the case of Vietnam', in C. Brundenius and B. Göransson (eds), *Reconstruction or Destruction? Science and Technology at Stake in Transitional Economies*, Hyderabad: Universities Press (India).
Tran Ngoc Ca (2007), 'Innovation systems in Vietnam: toward an innovation policy for competitiveness and sustainable development', *Journal of Science Policy and Research Management*, Japan Society for Science Policy and Research Management, Tokyo, **22**(2), 117–25.
Tran Ngoc Ca (2011), *Review of Vietnam's Innovation Policy*, Background report prepared for the joint OECD-World Bank Review of Vietnam's National Innovation System.
Tran Ngoc Ca (2013), 'STPI: from a project to the impact and some thoughts for the future', Note prepared for STPI project 40th year workshop, Lima, Peru, 5–7 August.

UIS (2015), UNESCO Institute of Statistics, Montreal.
UNESCO (1990), *World Directory of National Science and Technology Policy Making Bodies*, second edn, Science Policy Studies and Documents, No. 71, Paris: UNESCO.
UNESCO (2010), *Sistemas nacionales de ciencia, tecnología e innovación en América Latina y el Caribe*, ed. G. Lemarchand, Studies and Documents on Science Policy in Latin America and the Caribbean, Vol. 1, Montevideo, Uruguay: UNESCO.
Vu Cao Dam (2008), *Some Bright Milestones in the History of the Institute for Science Management*, Report to the national workshop, Hanoi: National Institute for Science and Technology Policy and Strategy Studues (NISTPASS).
Walters-Monteiro, M.H. (2010), 'Ciencia e Tecnología: Producao de Conhecimento em Mocambique entre 2003 a 2008', Master's Thesis, Universidade Federal do Rio Grande do Sul, Porto Alegre, Brazil.

4. The national innovation system in Bolivia and its relevance for development

Carlos Aguirre-Bastos, Javier Aliaga Lordeman, Ignacio Garrón Védia and Raúl Rubín de Célis Cedro

4.1 INTRODUCTION

For several years, large numbers of research projects have been executed in developing countries, with the aim to utilize the results for informing and influencing policy or to obtain social and economic innovations that could lead to economic growth and the improvement of the quality of life of their societies. Later, emphasis has been placed on the use of research results for inclusive development. A large number of studies have also been dedicated to study socio-economic processes, including science, technology and innovation (STI) to evaluate the effects of policy interventions.

Different studies (Aguirre-Bastos et al., 2010; Carden, 2009) show the difficulties that developing countries have in using research outputs for policy making. The former study shows that what is normally lacking is an explicit strategy for achieving policy influence, embedded across all projects as a standard consideration in their design. It is also verified that when policy analysis is carried out, it is typically concentrated on a set of issues related only to the project's core interest, without attention paid to the fact that often fundamentally important policies are spread across sectors, and rest on implicit rather than explicit assumptions about what promotes research and innovation.

Research outputs are also needed for social and economic innovations as elements of sustainable and inclusive development (Cozzens and Sutz, 2012; Sunkel and Infante, 2009). In developing countries, there are several examples of research impacting on local economies and of social innovations and minor or incremental innovations being produced not necessarily as a result of research but based on the ingenuity of actors. It is recognized that efforts along these lines need to be greatly enhanced if developing societies wish to successfully tackle social exclusion.

This chapter discusses the case of Bolivia. It analyses the contextual situation and experiences in both the periods previous to and post 2006 when a new government took office, changing the development paradigm that had been applied for decades.

Section 4.2 provides an overview of the present national political, social and economic context. Section 4.3 discusses the national innovation system, in particular developments in policy and governance, and provides an overview of the main actors and functions in the system, especially the university system where over 70 per cent of the research output is produced today. Section 4.4 describes policies and research outputs in the agricultural sector innovation system. Section 4.5 describes policies and research outputs in the health sector innovation system. Section 4.6 provides the results of a survey conducted in the research community on 'closing the loop'. The chapter closes with a discussion and conclusions.

4.2 THE POLITICAL, SOCIAL AND ECONOMIC CONTEXT

In 2006 the national development policy changed from a free market-capitalist model towards one based on the concept that the co-existence of different economic forms and made indigenous local communities (around 60 per cent of the total population of the country identify themselves as indigenous) recognize that their past non-capitalist economic relations were being substituted by a market logic, thus threatening their material and socio-cultural bases (Orozco Ramírez et al., 2006). In particular, the indigenous and syndical organizations of the country aspired to overcome the neoliberal economic system that had been implemented in the past decades and to reduce social exclusion. This vision has become the driving force for change since 2006 and has led to the adoption of a new Constitution in 2009 and an extensive normative reform.

The new Constitution defines Bolivia as a pluri-national State and it recognizes indigenous rights as a transversal dimension. Indigenous languages, symbols, ethical principles, land rights, political and judicial systems are enshrined in the text. Schilling-Vacaflor Del Carpio (2014) argues that such recognition makes this Constitution the most far-reaching of all other Latin American constitutions.

The new Constitution strengthens the already existing strong presidential system that has led in the past to concentration of power and lack of horizontal and vertical checks and balances. A report by the Economist

Intelligence Unit for the BBC[1] classifies the present political regime as a hybrid, a mixture between democratic and authoritarian. Under such a regime there is a danger that the rule of law weakens and that both the congressional and judicial branches of the State are not totally independent.

The *National Human Development Report* (UNDP, 2011) recognizes that one of the most relevant changes that has taken place in the past years is in the territorial reorganization of ethnic identities, a phenomenon that has allowed indigenous groups to realize that their territory, understood as a system of actors and social relations, defines common features and social distinctions that generate both inequity and possibilities for social mobility.

Change is also characterized by educational and occupational mobility across generations. Young people have more schooling compared to their parents, and recent educational achievements result in increasing mobility in the poorer households. Occupational mobility is also associated with changes in the economy and labour market structure.

In the past ten years the economy has grown at high rates. Gross national product (GNP) grew by 6.8 per cent in 2013 and it was expected that Bolivia would be the second fastest growing economy in Latin America in 2015. Since 2006 the economy has doubled in size. GNP per capita reached USD 2200 compared to USD 1200 five years earlier. Bolivia increased its exports to record highs, USD 12,500 million in 2013 up from USD 3000 million to 4000 million in the past decade, with gas representing more than 50 per cent of the total. The observed growth is the result of favourable international conditions for commodities and heavy investments made in the 1990s and early 2000s.

An important economic activity in Bolivia takes place in the urban informal sector. Between 2009 and 2010, 60 per cent of about 5 million economically active persons belonged to the sector. It is estimated that the sector varies from 72.2 per cent of the total employment in the city with the largest number of active workers to 57.6 per cent in the city with the lowest number. The sector contains workers who by their own decision belong to it, those family or apprenticeship workers that do not receive a salary and patrons and workers that undertake their activities in organizations with less than five employees (Evia et al., 2010). It is noted that these actors and their practices draw much from globalization and modernity in order to consolidate their style of economy.

The social context is characterized by decreasing poverty but persisting inequality. It is estimated (UNDP, 2014a) that Bolivia has seen the largest poverty decrease in Latin America of 32.2 per cent in the period 2000–12. At the same time, however, the country also has the highest percentage (16.9 per cent) increase in the size of the vulnerable population. These

Table 4.1 Human Development Index in Bolivia

	Life expectancy at birth (years)	Average schooling (years)	Expected schooling (years)	Income per capita (USD PPP 2005)	HDI
1980	52	4.5	9.4	3791	0.494
1990	55.8	6.4	11.3	2922	0.554
2000	63	7.4	13.7	3472	0.615
2010	66.3	9.2	13.5	4163	0.658
2011	66.6	9.2	13.5	4315	0.661
2012	66.9	9.2	13.5	4444	0.663
2013	67.3	9.2	13.2	5552	0.667

Note: PPP is purchasing power parity.

Source: UNDP (2014b) and Human Development Index Trends 1980–2013.

figures show that a decrease in poverty has corresponded in similar proportions to the increase in size of vulnerable and middle class populations. As seen in Chapter 2, Bolivia has also made some progress in reaching the Millennium Development Goals (MDGs), but has been slower than the other three countries involved in this study (Hailu and Tsukada, 2011). Social improvements are reflected in the Human Development Index (HDI) as shown in Table 4.1.

In spite of the observed improvements, inequality persists and has been very slow to decrease. Figure 4.1 shows the evolution of the GINI index in the period 1991–2012. In 2013, according to United Nations Development Programme (UNDP) data, the country is one of the nine Latin American countries that are among the 20 most unequal in the world (UNDP, 2013).

The observed social improvements were enhanced by new policies and reforms that since 2006 have taken place thanks to increased financial resources: *Bono Juancito Pinto* to promote and increase elementary school attendance; *Bono Juana Azurduy* to reduce child mortality rates; *Renta Dignidad* to economically support people 60 years of age or older not covered under any retirement plan; and the fuel subsidy. These policies have increased the government's public spending, as shown in Figure 4.2.

In light of the observed improvements there remain other grave challenges. The first is sustainability, as most growth is done through subsidies. A second challenge is the lack of productive diversification. The Strategic Research Programme (PIEB, 2011) has pointed out that the economy does not perform well due to the low factor productivity resulting from

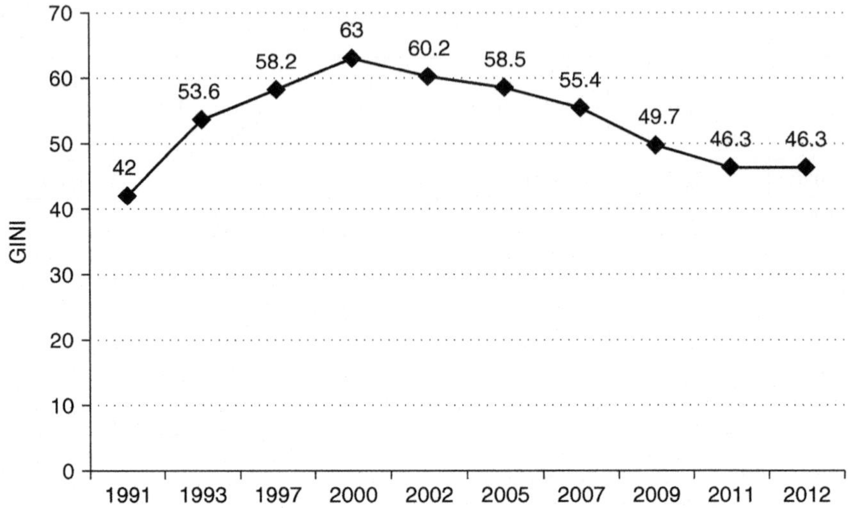

Source: UNDP (2014a).

Figure 4.1 Evolution of the GINI coefficient, 1991–2012

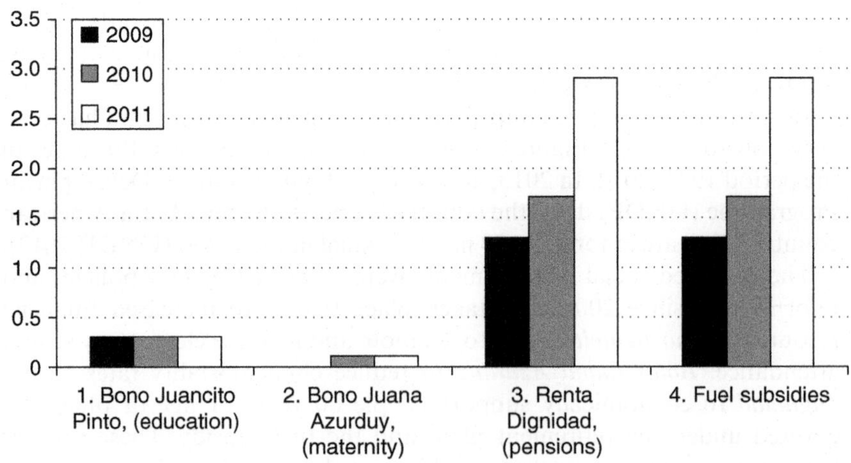

Source: UNDP (2013).

Figure 4.2 Main subsidies (as percentage of GNP)

institutional inefficiencies and weak political development. It is argued that the high dependence on natural resources impacts social actors, which in turn causes significant political conflict, impacting negatively on the economy. At the same time, little attention is paid to the linkages in the productive texture and the density of inter-industrial relations. Absent are stronger mechanisms to obtain productivity growth such as education and professional training, technology centre or university–industry linkages. Also influencing the performance of the economy is the existence of very weak clusters and value chains as well as lack of innovation.

4.3 THE NATIONAL INNOVATION SYSTEM IN BOLIVIA

4.3.1 Policy, Governance and Definition of the National Innovation System

In 1960, the National Academy of Sciences was established as an autonomous public institution in charge of the definition of science policy. Considering the need for creating a framework that would respond to broader social and economic demands, in 1977 the government established the National System for Science and Technology Development,[2] with a Directorate in the Ministry of Planning as its main executive organ.

From 1982 the political and economic crisis that affected the country isolated the Directorate from the Ministry's decisional processes and had many difficulties in leading the national science and technology (S&T) system (Aguirre-Bastos, 2001). This situation led in 1991 to the reorganization of the institutional framework by the establishment of the National Council for Science and Technology and an Executive Secretariat. The presidency of the Council was charged to the vice president of the Republic, thus providing strong political support.

In 1997, the framework was again changed as the Ministry of Education created the Vice Ministry for Higher Education, Science and Technology, transferring all duties of the former Executive Secretariat to this new organ. In 2006, the Vice Ministry of S&T (VMCT) was established under the Ministry of Development Planning, and in 2010, it was transferred to the Ministry of Education. The continuous changes in the institutional framework have affected the operation of the national innovation system.

The first S&T policy was drafted in early 1980 and in September 1981, guidelines were approved by government and a first national plan defined by 1982, but was never formally adopted; it did serve, however, as a guideline for institutions to define their own policies. In 1993 a national conference requested the formulation of a new plan, which was prepared by the National Academy of Sciences and the Executive Committee of the Bolivian University, but again never formally adopted.

In 1996, a Short Term Policy and Action Plan were adopted by the National Council for S&T. This was the first policy to be implemented and also the first that included innovation. One result of the Plan was the definition of a new strategy for the period 1997–2002 defined under the economic context of the time: 'Innovation for Competitiveness and Sustainable Development'. This approved strategy was never executed.

With the support of the Inter-American Development Bank, between 2003 and 2004, the Directorate of Science and Technology prepared the National Plan for Science, Technology and Innovation 2004–09. By the end of 2004, the Plan was formally approved by the six ministries overseeing the S&T area, but never executed.

In 2002 the Ministry of Economic Development defined a proposal for competitiveness policy and strategy (Velazco Reckling, 2002), which included science and innovation as a component (Aguirre-Bastos, 2002). The competitiveness agenda was only implemented in its first phases up to January 2006, when government changed.

In June 2001, Law 2209 'For Promotion of Science, Technology and Innovation' (STI) was promulgated as the first legal framework at that level. In 2010 the Vice Ministry of S&T proposed a reform to the Law based on the Constitution of 2009 that is still to be treated in Congress.

Article 103 of the new 2009 Constitution explicitly recognizes for the first time the importance of research, science and technology to support the development of the productive base and the implementation of strategies for incorporating knowledge and application of new information technologies.

The National Development Plan (NDP) 2006–11 (MPD, 2006) emphasizes a social vision and sets a full-fledged development agenda. It considers STI as a fundamental tool in the attainment of its social objectives and to change the primary export productive pattern, through the transformation processes of natural resources and the incorporation of new products in the market. Under such a vision, the Plan assigns technology and innovation a fundamental role for increased productivity and competitiveness.

Further, the Plan envisages that the role of STI will become operational through the organization of the Bolivian System of Innovation (SBI), and defines that SBI will develop technological based solutions to productive demand with the incorporation of knowledge to the product generation processes to satisfy quality, novelty, diversity levels and will also be assigned ecological and social validity. The latter implies that technology will not be applied indiscriminately but rather combine modern advances with ancestral, local and popular 'knowledges'[3] and wisdom (Carvajal, 2009; Carvajal and Albarracín, 2007).

The emphasis of government on ancestral knowledge as a basis for development comes from the fact that Bolivia, more than other countries in the region, still conserves an almost intact indigenous social life, so restoring traditional knowledge is considered not simply an anthropologic task but, rather, a way to rediscover more efficient techniques of production and management of the land. Thus, there is a development vision based on the belief that a dialogue between local and traditional knowledge, academy and enterprises, is possible. Such an approach is termed 'dialogue between knowledges'.

Bolivia has been leading a worldwide movement to introduce and advance the concept of ancestral 'knowledges' into a development strategy. This is thought to be an important step towards social inclusion of the 37 indigenous nations existing in the country. Later in this chapter, a more detailed description is made of the efforts and advances of the country in the treatment of this issue from the research and innovation perspective.

In 2013, government adopted a 'Patriotic Agenda' with the aim of reaching a set of specific goals by the year 2025, the country's 200th anniversary of independence. The Agenda contains 13 points, of which No. 4, 'Scientific and Technological Sovereignty with National Identity' sets guidelines and goals for the innovation system:

- Ownership and development of technology based on the convergence of knowledge in dialogue between science and local ancestral and community practices and knowledges for a given set of products in the framework of respect to 'mother earth'.
- To become a centre for technology innovation in nutritive and medicinal foods and with innovation centres in areas of production of high value native agricultural goods.
- To develop and strengthen knowledge and technologies for organic and conventional production of high yield products on the basis of the fusion of local, ancestral and modern knowledge guaranteeing abundant production of food and medicines.
- Recovery, development and strengthening of medical ancestral knowledge and practices in convergence with modern knowledge and practices and training and accreditation of professionals and highly specialized medical centres as well as the basis for the development of a natural, ecological and spiritual pharmaceutical industry.
- To increase and substantially improve the country's high level technology professionals, technicians, academicians, scientists and experts in different areas of knowledge, trained with the support

of the State, contributing with knowledge to the development and of 'living well' in harmony with 'mother earth'.

4.3.2 The National Innovation System

The proposal for the formal establishment of the Bolivian System of Innovation (SBI) (Carvajal, 2009) is derived from the NDP, and it follows a definition of innovation given as: '[T]he incorporation of knowledge to productive or social systems that generate new or better products, processes, and uses for the solution of local, regional and national problems, as an instrument that leads by itself to live well'.

The concept of the SBI is based on the need to articulate public sector led demand and is defined as: '[T]he set of interrelated and complementary actors that utilize science, technology and innovation in a coordinated and constructive way in the generation of integral solutions to productive, social and environmental problems with an approach of participative, socially balanced and sustainable development'.

In 2013, after the VMCT proposal for the definition of a national plan for STI (discussed below), the Bolivian System of Innovation was renamed the National Science, Technology and Innovation System. In this chapter, the term 'national innovation system' (NIS) is used to identify the latter. Figure 4.3 provides a glimpse on the major actors of the NIS.

Research infrastructure

Bolivia has been producing S&T indicators since the early 1990s drawn from existing databases. The collection and processing of information was organized following the recommendations of United Nations Educational, Scientific and Cultural Organization's (UNESCO) statistical procedures, as well as the Frascati Manual (Tellería, 1993).

In 2010 the VMCT conducted for the first time in the country a national research and development (R&D) survey. In the survey 304 R&D organizations were recognized, to which questionnaires were sent; 189 of them responded. A new online survey was conducted in mid 2012, in which the questionnaire was sent to 317 identified R&D organizations; 251 responded, thus greatly improving the response rate when compared to the previous survey. The main results of the latter survey are presented here.

Table 4.2 shows the number of research centres by sector and research area. The total in the table differs from the total number of centres that responded to the 2012 survey as in some of them there exists more than one research area. In what follows the university system (both public and private) is analysed in some detail, while the government R&D structure

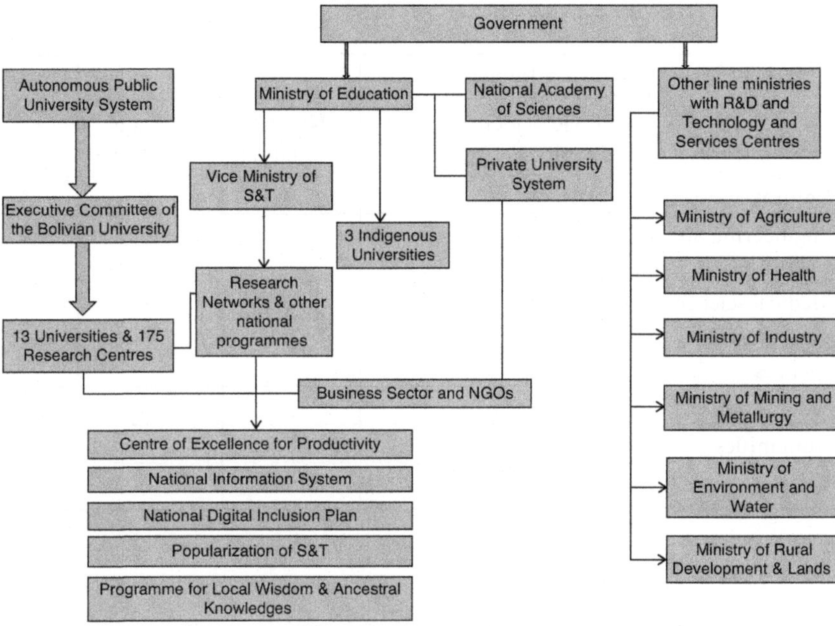

Note: NGO is non-governmental organization.

Source: Carlos Aguirre-Bastos.

Figure 4.3 Main actors of the national innovation system in Bolivia

is discussed in relation to the agriculture and health sector innovation systems, as these contain the largest number of public R&D centres in the country.

Table 4.3 shows the number of R&D and STI activities projects. Of the total number of these projects, 59 per cent were financed by the universities' budgets; 17 per cent by the contribution of the hydrocarbon direct tax; 16 per cent by international cooperation; and 8 per cent by sales of services.

Human resources in the national innovation system

Table 4.4 provides the number of personnel in research and other S&T activities. In this table as well as Table 4.5, which shows the number of researchers and fellowship holders by scientific area, fellowship holders are mainly MSc level students. In Bolivia, there are few Doctoral courses and as such PhD candidates should formally be the only ones to be included in the indicators (following the Frascati Manual).

Table 4.2 Number of research centres by research area

Research area	Public		Private		Total
	Public university	Government	NGOs/ foundations	Private universities	
Natural sciences	41	2	2	8	53
Engineering and technology	65	1	–	13	79
Medical sciences	22	1	2	9	34
Agricultural sciences	40	4	5	3	52
Social sciences	36	1	3	15	55
Humanities	6	–	–	5	11
Total	210	9	12	53	284

Source: VMCT database, www.cienciaytecnologia.gob.bo (accessed 19 February 2015).

Table 4.3 Number of projects in R&D and in S&T activities (2009–11)

Types of projects		2009	2011
R&D projects	Basic research	137	336
	Applied research	392	515
	Experimental development	71	49
	Total R&D	600	900
S&T activities	Technology innovation	40	94
	Transfer of results[a]	115	164
	Local knowledge	24	47
	Technology transfer[a]	67	41
	Total S&T activities	246	346
Total		846	1246

Note: a It is not clear from the data what is the difference between transfer of results and transfer of technology. It is assumed here that the former signifies the diffusion of research results by different documented channels.

Source: VMCT database, www.cienciaytecnologia.gob.bo (accessed 19 February 2015).

Table 4.4 Personnel in science and technology activities in 2011

Personnel	Time dedicated to R&D	Public		Private		Total
		Public universities	Government	NGOs/ foundations	Private universities	
Researchers	Headcount	1181	93	265	450	1989
	Full time	740	70	144	147	1101
	Part time	441	23	121	303	888
Fellowship holders	Headcount	351	6	26	135	518
	Full time	115	6	12	24	157
	Part time	236	–	14	111	361
Technicians	Headcount	239	8	68	127	442
	Full time	120	8	61	13	202
	Part time	119	–	7	114	240
Other support personnel	–	–	–	–	–	440
Other service personnel	–	–	–	–	–	172

Source: VMCT database, www.cienciaytecnologia.gob.bo (accessed 19 February 2015).

Table 4.5 Number of researchers and fellowship holders by scientific area in 2011

Scientific area	Researchers		Fellowship holders	
	Full time	Part time	Full time	Part time
Natural sciences	284	155	55	56
Engineering & technology	206	283	43	108
Medical sciences	116	113	16	69
Agricultural sciences	223	84	27	31
Social sciences	238	188	16	75
Humanities	34	65	–	22
Total	1101	888	157	361

Source: VMCT database, www.cienciaytecnologia.gob.bo (accessed 19 February 2015).

Table 4.6 Academic level by research area in 2011

Research area	Doctorate	Master	Bachelor	Technician	Others
Natural sciences	67	137	204	75	67
Engineering/ technology	48	182	254	25	131
Medical sciences	21	107	90	53	43
Agricultural sciences	23	96	169	26	52
Social sciences	64	165	184	51	53
Humanities	30	42	39	3	7
Total	253	728	940	233	363

Source: VMCT database, www.cienciaytecnologia.gob.bo (accessed 19 February 2015).

Table 4.6 shows the academic level of researchers and fellowship holders. As seen in the table, the number of PhD level researchers is very low. Further, in very few universities have policies for the recruitment of new researchers been adopted. This is, however, the case at the University of La Paz (UMSA) that grew from 224 researchers in 2001 to 467 in 2009. Of this total, 15 per cent have a Doctoral degree and 45 per cent a Master's degree. Of relevance here, OIM/SELA (2009) has studied in depth the brain-drain in Latin America, showing a rather dramatic situation. The analysis covered both the Organisation for Economic Co-operation (OECD) and the USA. In the particular case of Bolivia, it is found that 1200 Bolivians holding a PhD degree live in the USA, a high number indeed for a country with problems of critical mass.

Financial resources in the national innovation system

The 2009 survey of the VMCT collected financial information from only 30 per cent of the total sample, and data for 2010 and 2011 are quite contradictory, and for that reason are not reproduced here. From previous or alternative data estimates of expenditure in R&D as a function of GNP vary from 0.25 per cent (2009) to 0.11 per cent (2010). It is also estimated that 16 per cent of the total expenditure is provided by the private sector.

Expenditures in S&T activities in general have been extremely limited and are destined mainly to cover salaries and operational expenses. Expenditures in equipment, libraries, travel, all crucial for maintaining research capacity are almost non-existent. A national fund for research and innovation is still to be established.

Table 4.7 The Bolivian higher education system: number of institutions and organizations (2005–10)

Higher education institutions	2005	2006	2007	2008	2009	2010
Universities	67	71	77	84	85	85
Other higher education institutions	129	152	187	254	302	313
Total	388	443	524	679	788	816

Source: Arauco and Gallardo (2011).

There are some sector funds normally accessed by public competitions that have provided important funding for some sectors. This is the case of PIEB for social research and (former) SIBTA for agricultural projects, and funds for environmental research and protected areas.

Bolivia does not have a venture capital system for technology and innovation, but holds high potential in its microfinance system. The Global Competitiveness Index places the country in the 30th rank worldwide (WEF, 2014) in financial market development. It is yet to be analysed in some detail how to utilize this system for technology and innovation.

The higher education system

Universities in the higher education system The production of highly skilled human capital is certainly a key contribution that the higher education system makes to the NIS. In the system the role of Bolivian universities is paramount[4] as these institutions not only provide key human resources but also produce more than 70 per cent of the total research output.

The number of higher education institutions shows important growth after 2005 (Tables 4.7 and 4.8). The creation of three universities in 2007 to facilitate the access of indigenous peoples marks a departure from the traditional method of creation of higher education institutions. The main objectives of these universities is reconstructing native identities, develop scientific knowledge, 'knowledges' and technology, oriented by community criteria under the principles of complementarity, cooperative work, individual and collective responsibility and equilibrium with nature.

Table 4.9 shows the number of students, staff and graduates in the university system since 2000. It can be seen that there has been a large increase in the number of students since 2000, in both public and private universities. Also seen is a tendency that staff upgraded their academic level; between 2005 and 2007, 30 per cent of academic staff had obtained

Table 4.8 Number of public and private universities (2005–10)

Universities	2005	2006	2007	2008	2009	2010
Public autonomous universities	11	11	11	11	11	11
Universities under special regime	2	3	3	3	3	4[a]
Universities for indigenous groups				3	3	3
Total number of public universities	13	14	14	17	17	17
Private universities	40	41	42	42	42	42
Sub-seats of private universities	14	16	21	25	26	26
Total private universities	54	57	63	67	68	68
Total number of universities	67	71	77	84	85	85

Note: a According to CEUB (2012), there are four (not three as noted by these authors) universities under special regime: the Police University, the Andean University, the Catholic University and the Military School of Engineering.

Source: Arauco and Gallardo (2011) and CEUB (2012, p. 1).

Table 4.9 Students, staff and graduates in Bolivian universities

	2000	2005	2010	2011	2012	2013
Number of students in public universities	213,291	259,221	333,210	366,036	397,939	366,251
Number of students in private universities	44,357	72,307	91,532	103,863		
Teaching staff in public universities	7,869	8,329	11,798	12,212	13,060	
Teaching staff in private universities	4,103	7,185	7,504			
Number of graduates from public universities	13,614	5,658	17,987	20,162	19,990	
Number of graduates in private universities		4,929	3,810			

Source: National Statistics Institute, www.ine.gob.bo (accessed 31 January 2015).

Table 4.10 Number of registered students and title granted in postgraduate studies in the public university system (2005–10)

Year	Number of registered students	Number of titles granted	
		Number	%
2005	7929	2559	32.3
2006	8055	2124	26.4
2007	8475	1958	23.1
2008	10,256	2470	24.1
2009	10,661	2598	24.4
2010	9539	1586	16.6

Source: Arauco and Gallardo (2011).

a postgraduate degree, while this percentage increased to around 60 per cent by 2014.[5]

The postgraduate system is still small when considering the number of universities. In 2011 the total number of postgraduate courses offered in the university system was two 2 at the Doctoral level and 61 at the Master's level. There are other modalities in postgraduate education that have been used, in particular, the so-called 'diploma' courses that do not carry any research activity and only provide a diploma for attendance. Of this modality in 2011, 115 courses were offered. The number of students attending postgraduate courses is shown in Table 4.10. It is important to highlight that only around 24–30 per cent of students attending postgraduate studies actually receive their diplomas.

Governance of the university system in Bolivia is determined by the new Constitution of 2009 and the Educational Law of 2010. These instruments recognize: (1) autonomous public universities, congregated around the Executive Committee of the Bolivian University (CEUB); (2) private universities under the supervision of the Ministry of Education, making up the National Association of Private Universities; (3) indigenous universities as public organizations under the tuition of community boards; (4) special regime universities as public bodies under the tuition of their boards, supervised by the Ministry of Education.

Expenditure of the university system represents around 2 per cent of the GNP. The main financial resources made available to the public university system have increased from USD 172 million in 2005 to USD 445 in 2012 (Ministry of Economy and Finance, 2012). An important point to be made here is that support of the system is contingent to tax collection and the indirect tax on hydrocarbons. In 2012 public universities received an additional USD 31 million because of higher tax collection and higher international oil prices.

Research in the university system Public universities are, by and large, the main actors in R&D in Bolivia. The number of research centres in the public universities of the system is 175, of which 58 per cent are concentrated in the three main cities (La Paz, Cochabamba and Santa Cruz) – of these 41 are in technology and engineering; 38 in agricultural sciences; 32 in social sciences; 28 in pure and natural sciences; 27 in health sciences and 9 in the humanities. Several have received international recognition for their contribution to the generation of knowledge.

The Committee of the Bolivian University has made an in-depth analysis of the situation of public universities as part of the STI system (CEUB, 2012) as a first necessary step in the adoption of its National Strategy for S&T 2012–15. It is recognized at the onset that the public university system is yet to establish a reliable indicator and statistical system.

CEUB points out that the university structures that support the management of S&T, excluding the two largest universities, UMSA (La Paz) and UMSS (Cochabamba) are weak and unarticulated, with reduced personnel. Several universities have drawn strategic plans for research, which take years to be approved by the university governance system.

The present management structures in the universities have not been able to overcome the weak linkages between the generators of knowledge and the demand side, and there is a set of constraints that needs to be overcome, among which are the absence of processes to value the protection of intellectual property and lack of national legislation for technology transfer. Ahead of this situation, in 2010 the II National University Congress decided to create 'Linkage Units'. At the time only UMSS operated a Technology Transference Unit, successfully executing linkage programmes.

The financing of S&T activities and research in the university system represents only 7 per cent of the total university budget. As is the case of the whole STI system, funds from abroad (estimated to be at least 60 per cent) constitute a key component of financing. The direct tax on hydrocarbon has represented growing support for R&D since 2005, little less than USD 7.0 million per year for all the system. The excessive bureaucratic procedures for using these funds have discouraged researchers to apply.

When research projects are individually analysed, it can be seen that public universities have been working on issues dealing with the main social and economic problems of the country for years, and several achievements can be singled out in the areas of biotechnology, genetics, energy and others. In some cases, university research has provided assistance to vulnerable groups by carrying out research to improve health practices or environmental management. However, a radical cultural change is urgently needed at the universities if they are to be able to effectively support sustained and inclusive development. In particular, the priority and incentives given to quality research and research based education must be increased relative to the present dominant emphasis on traditional teaching.

Thulstrup et al. (2006) have made an in-depth evaluation of the SAREC/SIDA funded research activities in the two largest public universities (UMSS and UMSA) and have provided an overview of the situation in the whole system. Their main conclusions are:

- In theory, the relevance of the S&T project themes is generally good, in the form of a potential, direct impact on social, environmental or economic problems.
- Projects have realistic opportunities for successful outcomes, both in terms of research training, academic research results, and within a wide range of real applications and development contributions.
- Effectiveness is generally quite good with respect to research and research training, but low with regard to academic communication (including publishing) and practical applications of the research results and skills.
- The impact is mainly of a less practical nature but formal, academic qualifications of the staff are improving considerably. Societal impacts are so far quite limited, although it seems likely that several of the projects may increase their impact on development and poverty reduction considerably, both directly and indirectly.
- Important numbers of PhD and MSc students are returning; laboratories have been upgraded significantly and are generally successful; low-cost and local solutions have been cleverly used; research funds are changing the culture through their demand for well-documented proposals; extensive cooperation with foreign advisers of graduate students exists.

In spite of the positive signs, the study of Thulstrup et al. also points out problems:

- In policy and management: policies are not institutionalized; academic structures are weak; policy does not have continuity; no policy for training researchers exists; lack of linkages between national and university policies; application and dissemination of knowledge is insufficient.

Table 4.11 Evaluation of the S&T University Development Strategy 2007–10

Objectives	Mean value[a]
1. Make tangible the contribution of the university to the social and economic sustainable development of regions and the country, conducting research and innovation activities that contribute to improve the quality of life of peoples, increase productivity and competitiveness of the economy, generate employment and improve the quality of higher education.	2.57
2. Make compatible the productive, prefecture and municipalities and other sectors' needs with the interests and academic values of the Bolivian University, so that their resources and infrastructure are added and complemented to attain sustained scientific and technological development.	3.06
3. Develop a normative framework and an institutional operational environment to increase the production of STI, coherent with the demands of productive, prefectures, municipalities sectors and other development organizations.	2.34
Mean valuation of the Strategy	2.66

Note: a Scale: 1 objectives not accomplished; 5 excellent accomplishment.

Source: CEUB (2012).

- In dissemination of research: absence of a publications culture; very few Bolivian publications are international or are indexed; little quality control of most local publications; insufficient diffusion inside the academic community and outside.
- In the future of research: absence of research training; university plans produced without the participation of active researchers; absence of adequate infrastructure; insufficient administrative and financial structures.
- Of research as inputs to the quality of education: absence of an academic model centred on research; absence of a research administration culture; insufficient contribution of research to development; low priority of STI.

CEUB has evaluated the S&T University Development Strategy 2007–10 and Table 4.11 reproduces the main results. It should be noted that many objectives are those needed to be fulfilled to 'close the loop'. The result of the evaluation is interpreted as showing the difficulties in managing research inside the university environment.

On the basis of the results of the evaluation, a SWOT analysis was

Table 4.12 SWOT analysis of the Bolivian university system

Strengths	Weaknesses
1. The CEUB groups the largest and most qualified group of researchers in the country. 2. The CEUB has the largest infrastructure. 3. Growing use of information and communications technologies (ICTs). 4. Presence of authorities engaged in improving the environment for research and innovation. 5. Growing use of the direct hydrocarbon tax for R&D. 6. Established curricular development policy. 7. Human resources trained in management and transfer of research outputs.	1. Simple and beauracratic structures for managing S&T. 2. Insufficient financial resources. 3. Non-existent or poor critical masses in certain areas. 4. R&D centres with deficient infrastructure. 5. Weak linkage with socio-economic sectors. 6. Insufficient diffusion of existing services. 7. Lack of knowledge of demand of productive sectors. 8. Incipient information and communication system. 9. The largest human resources and material infrastructure is concentrated in three cities.
Opportunities	**Threats**
1. Participation in regional and national plans. 2. Commercialize services and advisory tasks. 3. Demand and expectation for university services. 4. Agreements at national and international levels. 5. Possibilities to obtain financial resources for research. 6. Possibilities to integrate national and international information networks. 7. Generate a change in attitude in the context of the national innovation process. 8. Process of certification and accreditation at national and international levels of existing laboratories.	1. Postponement in the generation of laws and decrees that favour development of STI. 2. Decrease of direct tax on hydrocarbons income. 3. Brain-drain at the postgraduate level.

Source: CEUB (2012).

produced (Table 4.12) illustrating the conditions under which research takes place in the public university system.

Based on the evaluation of the previous Plan and the SWOT, a new Strategy for 2015 has been defined and adopted by the university system. The Strategy defines a set of 12 policy objectives under three broad strategic objectives, as given below, which are similar to those in the previous Plan. It defines specific goals and actions to reach the set objectives. It also provides guidelines to evaluate the Plan during its execution and conclusion. No financial resources are estimated for the Plan's execution.

a. To promote the contribution of the university to the social and economic development of the country and regions, through research and innovation.
b. To attain a sustained scientific and technological development making compatible the needs of productive sectors, departmental governments, municipalities and other institutions, with the academic interests and values of the Bolivian University, so that resources and infrastructure in their fields of action are complemented and added.
c. To increase the production of STI in the university, through a convenient normative framework and institutional environment.

STI in the business sector

The efforts of the business sector to undertake research and innovation activities have not been measured formally. The country has not conducted innovation surveys based on the Oslo or Bogota Manuals. There are few specific case studies that provide a vision of innovation that take place in different sectors.

In the segment of large enterprises there are several that have attained technology development and innovations of value; this is particularly the case for the textile industry, oils and fats, and leather, which have become exporters to the USA and other markets. Alänge et al. (2004) have analysed local innovation and cluster activities in Bolivia, showing that there are many weaknesses in the innovation system, but there are examples of local innovation in communities (eco-tourism), entrepreneurial companies, and there are relationships with suppliers in several cases.

In recent years, many private sector associations have been more active in promoting technology and innovation development within their associates. As an example, in 2009 the Departmental Chamber of Industry and one public (UMSS) and one private (UPB) university agreed to create the Foundation for Research and Innovation, with the objective of linking enterprises with universities, to promote competitiveness of the business sector. It is expected that the Foundation will generate spaces for the actors in the regional innovation system to identify technology supply and demand.

A study by Aguirre-Bastos (1997) conducted on small and medium-sized enterprises (SMEs) through an extensive national survey showed that there are several weaknesses in enterprises for them to actively participate in the NIS: lack of consciousness and sensibility of managers towards research and innovation; little value given to human capital, SMEs do not invest in training or motivate their staff; scarce enterprise cooperation as there is no cooperation mentality with other agents of the innovation system.

In spite of this situation, it is also observed that several SMEs have an organizational dynamic that shows a tendency towards using quality control and the adoption of programmes for continuous improvement. These are developments in organizational innovations that can contribute to improve competitiveness of firms.

The government has manifested interest in the sector by creating the Ministry of Productive Development and Plural Economy. In addition, the Bank for Productive Development has been established, together with other mechanisms, for training, market analyses and so on and to provide loans and assistance.

In the business context, a number of institutions and organizations in Bolivia have been created or strengthened to service production since the early 1990s. Examples are the Bolivian Institute for Standards and Quality and the National System for Quality, Metrology, Accreditation and Certification. Also in this period the Bolivian Office for Accreditation and the National Service for Intellectual Property were reorganized.

Research and innovation networks

There are at present 14 research networks operating in the country, encompassing a broad range of scientific fields and multidisciplinary topics (Table 4.13). Because of its socio-political importance in the country, the Network of 'Knowledges of the indigenous, originary, peasant and Afro-Bolivian peoples' is described in more detail.

The Network of 'Knowledges of the indigenous, originary, peasant and Afro-Bolivian peoples' has been organized on the premise that in the academic environment, knowledge is generated based on the scientific method and that in the rural environments knowledge and 'knowledges' are generated based on daily life, practice and oral transmission. It is thus logical to think that from these forms of knowledge production, when combined and complementing each other, it is possible to build new knowledge. This process of symmetric construction and horizontal complementation of knowledge and 'knowledges' is termed the 'technology based innovation process based on the dialog of knowledges' (VMCT, 2012).

Table 4.13 R&D networks in Bolivia

Network name	Network name	Network name
1. Food 2. Biodiversity 3. Forests 4. Behavioural sciences 5. Nuclear energy 6. Renewable energies 7. Tech base incubators 8. Hydro resources	9. Environmental remediation 10. Knowledges of the indigenous, originary, peasant and Afro-Bolivian peoples 11. Information and communication technologies	12. Science, communication and culture 13. Geomatics and Earth's observation 14. Palaeontology

Source: VMCT, www.cienciaytecnologia.gob.bo (accessed 31 January 2015).

In Bolivia, ancestral technologies have an ecologic character that is manifested in harmonic development with the environment, resulting in equilibrium between society and its habitat. Knowledge for survival has been retained and conserved in the different communities. The recovery and preservation of such knowledge makes the communities resilient and their development sustainable. Thus, the combination and complementarities of ancestral and modern knowledge becomes a social demand that is expected to lead the construction of new science paradigms and the development of new educational and research programmes.

In this context, the network promotes the development and strengthening of STI through the recovery, revitalization, appropriation and register of ancestral 'knowledges' and local knowledge. This objective is obtained by articulation of the different actors involved in the process. At present there are ten institutions that belong to the network, of which three are universities, five are NGOs, one is a government agency (the Ministry of Rural development and Lands) and one is the Confederation of Indigenous Peoples of Bolivia.

The network has been able to design research methodologies, new educational curricula and build an information base of importance. The research outputs of the network in relation to the agricultural sector are discussed further below.

Besides the work being undertaken by the network's institutions, there are several other efforts to combine modern and ancestral knowledge for development. Some are detailed when analysing the agriculture and health sector innovation system.

Research outputs

'Closing the loop' depends strongly on how the scientific community transmits the results of its work. Ideally, research results must be made available for:

- the international scientific community
- the local users such as industry, local communities and so on
- policy makers, interest groups and organizations that require scientific opinion in respect of specific issues for the definition of policies.

In Bolivia, dissemination of research results fails in all categories, in spite of existing talent, the value of research results and the efforts to publish in local journals.

Many scientific publications are made in local journals, but few are indexed in international databases due to lack of criteria to define the level of advancement and development of published journals and the lack of procedures, norms and national criteria to define the publication index (Moraes, 2009). On the other hand, many journals lack editorial committees. It should be noted that some countries are managing to merge a large number of more or less ineffective local journals into a small number of strong, international journals.

The Vice Ministry's surveys identified scientific and technological publications (books and journals), independent of indexation, periodicity or refereeing criteria. It found that in 2011 there were 183 electronic books published in all scientific areas, 390 physical books and 468 journals. Of the large number of existing journals, only seven in the social sciences, three in S&T and three in the medical sciences comply with international standards regarding form and content. The Scielo Platform comprises 15 journals that conforms to defined quality criteria. Of these, ten journals belong to the public university system (of which four are from the Catholic University, four from UMSA and two from UMSS). One journal (*Ecology in Bolivia*) is indexed in Latindex, Dialnet, Periodica and LILACS.

SCImago Lab (2015) in its publications and journal rankings has identified for Bolivia, for the period 1996–2013, 2874 documents, which received an average of 13.79 citations per document, placing its H Impact Index at 71. When compared with the other countries included in this book (Mozambique, 1762 documents, H Index 60; Tanzania, 9170 documents, H Index 102 and Vietnam 20,460 documents, H Index 122), Bolivia shows difficulties in producing more impact publications. When the number of publications is related to population, Bolivia stands out from the four countries as the larger producer with around 287 per million inhabitants.

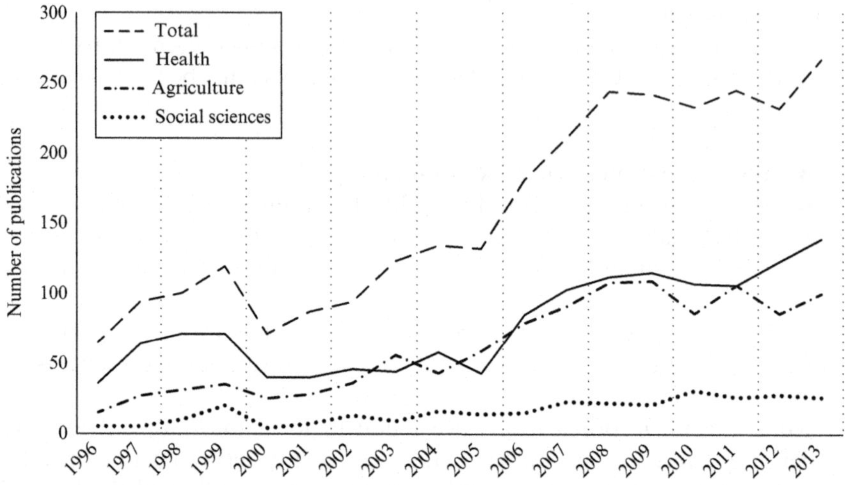

Note: Total means the total number of publications in all sciences; social sciences does not include economics, of which there is a total of 33.

Source: SCImago (2015).

Figure 4.4 Evolution in the number of publications per priority area

For comparison, in the same period, Sweden produced 417,156 documents with an H Index of 567.

Figure 4.4 shows the evolution of publications in Bolivia for agriculture, health and social sciences, demonstrating that these research outputs do follow the two important national priorities.

Lemarchand (2012) has conducted a bibliometric analysis for Ibero American publications between 1973 and 2012 in the SCI, SSCI and A&HCI databases. In his rankings Bolivia appears in 17th place out of 18 countries with 2806 publications compared to 703,031 of Spain and 363,581 of Brazil, which are the largest in this region.

Moreover, the number of local patents continues to be very low in relation to the existing potential to register. Table 4.14 shows the indicators for the years 2010 and 2011.

4.3.3 The National STI Development Plan

A priority line of work of the VMCT today is the implementation of the National STI Development Plan adopted in 2013 (VMCT, 2013). The Plan correctly recognizes previous efforts to build up an S&T system

Table 4.14 Patent requested and granted at the National Patent Office

	Patent requests		Patents granted	
	Residents	Non-residents	Residents	Non-residents
2010	80	333	5	67
2011	118	322	48	66

Source: VMCT database, www.cienciaytecnologia.gob.bo (accessed 19 February 2015).

and concludes that the deficiencies encountered in them stem from the historical perspective of the absence of State policies. It recognizes that the limitations of the existing NIS are both structural and strategic. The Plan identifies and recognizes the research work being undertaken by nearly 2000 researchers in 251 research centres, a large majority of which are located in the public university system, and the few research activities being undertaken in the private sector.

The Plan acknowledges the key role of universities and the need to strengthen postgraduate education and democratization of knowledge through diffusion programmes and projects. In the former case it is recognized that the present postgraduate supply of professionalizing Master level courses (a modality invading Doctoral courses) is insufficient to generate STI capacities able to respond to social and economic demands.

The Plan considers that five problems must be addressed with high priority: (1) formulation and implementation of normative instruments, including a new Law for S&T; (2) inject in a sustained manner the necessary financial resources into the system; (3) diffuse information at all levels so that research outputs reach decision makers, researchers follow national policy guidelines and the demand for technology is satisfied; (4) develop in time and apply with due frequency evaluation and monitoring instruments to assess and provide guidelines; and (5) generate human talent able to respond to societal demands.

As a result of the process of consultation of the Plan with the innovation system's stakeholders, priorities have been set for a first implementation phase between 2014 and 2019 (a second will go up to 2025) on the training of human resources, particularly postgraduate education as a horizontal priority. A set of 19 lines of action and 64 sector programmes have been identified as priorities under eight key sectors: health, agriculture, industry, local wisdom and ancestral knowledge, natural resources and environment, energy, mining.

It is quite interesting to note that the Plan does not contain explicit reference to research in the social sciences and the humanities. In both cases there are important research activities in the country and, relevant to this chapter, the role played by the Strategic Research Programme, PIEB, has been crucial to advance knowledge and inform and impact on policy making in the country. There are other key contributions to social research in several universities as well, for example, the Catholic University and the larger universities in the public system, as well as in several NGOs.

4.4 THE AGRICULTURE SECTOR INNOVATION SYSTEM

4.4.1 Overview

Between 1949 and 1966, at the instruction of the Inter-American Agricultural Service (established in 1948), 11 agricultural and three demonstrative livestock experimental stations were created. It is recognized that this is the period when research and innovation reached high standards. Efforts in this period were strongly influenced by the green revolution concepts, and were dedicated to study the most appropriate conditions for the production of a large number of products. Research results were transmitted by an able extension service also established at the time (Cardozo Gonzales, 2001; Flores, 2003).

With the creation in 1975 of the Bolivian Institute for Agricultural Technology (IBTA), research and extension was taken over principally by the State. At this time the traditional model of research institutes was consolidated following the green revolution model and the industrialization of agriculture. It is widely considered that the intensive agricultural model adopted was not adequate to the (dry) western or Andean (mountainous) production zones, as it required irrigation, but was more successful in the more resource-rich eastern lands.

In 1990 a process of institutional reform was undertaken, as part of the structural adjustment, modernization and reduction of the State process. In this period the role of international cooperation was fundamental for the continuation of research, but the extension services were completely dismantled. The process of institutional reform was completed only in 2000 with the creation of the Bolivian Agriculture Technology System (SIBTA), whose main tasks was to promote and develop technology, forestry and agro-industrial innovation, and to articulate the demand of actors in agro-productive value chains. SIBTA operated through four foundations in the main national macro-regions, and defined two sets of projects: Projects for Applied Technological Innovation (PITA) and National Strategic Innovation Projects (PIEN).

According to an impact evaluation of 24 PITA projects (Lema et al., 2006, cited in Delgado Burgoa and Escobar Vásquez, 2009), there have been positive effects in income and the mean return rate. It has also been noted that there was an induced demand by the suppliers (62 per cent), weak participation of demand in project design (66 per cent) and weak benefit for vulnerable groups (40 per cent). Public universities were excluded from projects as potential suppliers and priority was given to consulting firms. In 2004 an assessment of the R&D capacities of the existing 69 agricultural centres (nine private, one mixed and 59 public, including university centres) showed that only eight had adequate capacity, 41 moderate capacity and 20 severe limitations to undertake R&D.

One example of private innovation efforts in agriculture that has been undertaken since 2000 is Fundación Valles (2010),[6] whose membership is composed of 21 private and public institutions. The technology innovation programme is the main axis of work. It is strongly based on a participatory approach and has paid attention to value chains, based on long-term programmes. This approach has been effective in the identification of critical factors that limit the development of agri-business and that limit the adequate economic development of the small producers.

To December 2011, the Foundation had executed 12 programmes, which include 136 projects (finished) and 14 ongoing, benefiting more than 58,184 families grouped in 122 organizations and small enterprises. The interventions are mainly centred in the diffusion of existing technology based innovations. Through its projects, it has been possible to develop suppliers of inputs and services, and offer access for small producers to export markets.

Of importance in the private sector is the presence of the research and extension services of the Patiño Foundation[7] established in 1958, which contribute to the development of education, culture, research, health, hygiene, nutrition, agriculture and ecology. The Foundation operates through various centres, owned by the Patiño University Foundation created in 1931. The latter operates in Europe, diffusing the richness and diversity of Latin American cultures and maintains a university centre in Geneva that provides scholarships to deserving Bolivian students in Swiss universities.

In the area of agriculture, the Foundation operates three organizations: the Centre for Phytoecogenetic Research; the Model Farm; and the Seed Centre. The research and innovation effort concentrates on specific plant varieties (and livestock), and in collaboration with the Seed Centre it diffuses its products and information on use through training courses, demonstrative plots and preparation of technical manuals.

The production of high quality genetic and phyto-sanitary seeds is done through contracts with farmers and the Model Farm, to whom basic seeds are provided and technical assistance provided through the production cycle. The Seed Centre produces over 200,000 kg of certified seed of improved varieties for sale to small producers. Nearly 40,000 of the latter cultivate these seeds. The increase in income of small farmers that use the Centre's varieties is estimated at USD 10 million per year.

One of the most important and evident impact of the model pursued for several decades is that of the Andean grain quinoa; the present commercial varieties have been obtained through extensive research efforts on behalf of IBTA and other actors of the sector innovation system. In 2011, the FAUTAPO Foundation prepared full documentation to back and register the 'denomination of origin' of the Quinoa Real species, which is today widely commercialized worldwide.

The research efforts allowed 'closing the loop' in this specific case. Antonio (2011) has analysed both the potential and existing policy challenges for the development of a sustained quinoa based agro-industry as described in Box 4.1.

Other research efforts by both public and private non-university agents with positive results can be identified from the early 2000s. Examples are those of the PROINPA that makes part of the system of natural genetic resources, using molecular markers to evaluate diversity in their culture collections, and working in programmes to improve corn, quinoa and other species.

Universities have also contributed extensively to the development of the sector system, for example, the University of Santa Cruz (UAGRM), which has worked on commercial fruits; the Military University on micro propagation of flowers; and UMSS in micro propagation of fruits.

Hartwich and Jansen (2007) have extensively analysed the role of government in agricultural innovation in Bolivia. In their analysis the authors pointed out that governments in developing countries, when trying to manage complex innovation processes involving many and different actors, sometimes find it difficult to design effective interventions and therefore end up supporting and managing only the public research and extension organizations that directly depend upon them.

This has not been the case for Bolivia, however. The study of Hartwich and Jansen found that despite a number of weaknesses related to the design of the system and the government's limited commitment, the regional foundations were able to effectively identify the demands of small farmers, set priorities and provide transparency and accountability with regard to funding and decision making. The key empirical findings of the study that may be considered as lessons learned may be summarised as:

BOX 4.1 SUCCESS AND CHALLENGES OF A SUSTAINABLE QUINOA BASED AGRO-INDUSTRY

The value of Bolivian quinoa exports has risen from a total of USD 2 million in 1999, representing 1500 tons, to more than USD 46 million in 2010, or about 15,400 tons. Exports have attracted international customers and exporters willing to buy the grain at higher prices. In 2010 there were 23 major quinoa exporters, all part of the National Association of Quinoa Producers (ANAPQUI) whose combined revenue that year was about USD 42 million. The analysis made shows that if production and exports are to be sustainable, the agricultural practices must be both ecologically and socially sound. Producers face technical difficulties and environmental risks of importance, and there are impacts on the social and environmental capital of the regional economy that need very special attention.

The policy problem in the long term is that the expanded growth of the industry represents both a great opportunity and potentially a serious threat to the environment and local nutritional levels among the poorer communities. The high prices and worldwide demand have put tremendous pressure on production, and this pressure has encouraged poor environmental practices as well as illegal trafficking.

To face the challenge, Antonio's study calls for ANAPQUI to promote and assist farmers to implement sound environmental management practices, focusing on how to increase production by increasing the yield instead of expanding the agricultural surface; government should solve trade issues and help make the industry more competitive. One step in this direction has been the quinoa law approved in March 2011 that establishes property rights over the variety *Royal Quinoa*, and provides for other means of support. But government needs to improve the business environment, as well as trade conditions and infrastructure. Antonio's study also recommends that stronger partnerships are required between ANAPQUI and the government. For example the latter promoting local nutrition programmes by providing quinoa to the most vulnerable groups, incorporating the grain into the school feeding programme run by the Zero Malnutrition Programme.

Source: Antonio (2011).

a. The research and technology transfer programme that was carried out by SIBTA was only one component of an innovation system and thus there remains other complementary functions that need to be contemplated with priority by government to foster innovation, focusing on overall planning on the macro level and on policy analysis, the setting of consultation platforms, supporting the building of innovation networks and setting up specific funding mechanisms.
b. Setting up decentralized semi-autonomous agencies that administer funds and design innovation projects is not enough to allow participation of local producer organizations and technology providers. It is necessary that more participation be promoted by the adoption

of special rules and incentives to collaborate and the special efforts of all involved, and eventually further decentralization on the regional level.

c. Weak central governance, rather than a decentralized structure or the delegation of too much power, have prevented governments from taking a more active role in governing their innovation systems.
d. Being responsive to the demands of farmers does not necessarily mean creating the best technical solutions. Generating innovations requires the participation of all actors in the innovation system, as well as analysis and identification of technological and market opportunities.
e. Policy makers should foster in-depth analysis of farmers' demands on the local level through decentralized organizations, which simultaneously help to orient these demands to where technological and market opportunities lie.

4.4.2 New Developments in Agriculture Policy

Considering that the existing agriculture production model had limited impacts, several authors (for example, Delgado Burgoa and Escobar Vásquez, 2009) have indicated that the model does not value the local 'knowledges' and wisdom of indigenous people, nor the wealth of the existing biodiversity, as the model introduced only modern technologies to replace the ancestral systems.

Under the above premise, in 2006 the new government established the National Institute for Innovation in Agricultural and Forestry Innovation (INIAF), whose mission is defined as 'the sole research instance accredited in scientific exchange activities at the national and international levels, related to agriculture and forestry issues'. The Decree creating INIAF points out that 'research must be undertaken from the State (in association with universities and the private sector) for all rural producers'.

In the view of Delgado Burgoa and Escobar Vásquez, the innovation vision taken by INIAF is an adequate alternative as it orients to the prioritization to food security and autonomy, and incorporates local 'knowledges' in an integral development vision. As it takes issue with a view of innovation oriented principally to competitiveness in the market, value chains and productive specialization, it is considered that it does form the basis for the construction of sustainable management of natural resources and the promotion of biodiversity allowing endogenous sustainable development. Table 4.15 provides a comparative analysis of the traditional and alternative innovation approaches.

From their analysis, Delgado Burgoa and Escobar Vásquez have defined the agriculture sector innovation as: 'the participative generation or

Table 4.15 Comparative analysis of the traditional and alternative innovation approaches

Category	Conventional	Alternative
Approach	Neoliberal approach that emphasizes the market and reduces the role of the State giving emphasis to enterprise development and industrialized agriculture for exports. Does not recognize the local 'knowledges' and wisdom of local communities that have had a fundamental role in food security in the country.	Approach for a sustained endogenous development triggered by a dialogue among 'knowledges' (modern scientific western knowledge, local 'knowledges' and wisdom of local communities) emphasizing the complementarily of industrialized agriculture and that of small producers.
For what?	Greater competitiveness in the market ahead of the opening of new markets, climate risks and diminishing soil productivity.	Guarantee food security and autonomy with greater competitiveness in the local and external markets, ahead of climate change and diminishing soil productivity.
How	Disciplinary and interdisciplinary development in the framework of western modern science and technology, through conventional methodologies and vertical trends. Prioritization of vertical technology transfer.	Transdisciplinary, revalue of local 'knowledges' and wisdom of local original communities, as opening of an inter-scientific dialogue. Participatory research is revalued and the participatory technology development are usual methodologies in a perspective of a multi- and inter-methodological perspective.
Where	Efforts are prioritized in experimental stations with some efforts of participatory research of instrumental character and the transfer of technology.	Research and innovation are prioritized as part of a dialogue of 'knowledges' between local actors and professional technicians. They are carried out in the field (producers' lots, communities, production centres etc.) with the participation of local actors.

Source: Delgado Burgoa and Escobar Vásquez (2009).

re-creation of a knowledge, technology, product or process, in the framework of sustained management of the whole of the productive system, through the re-value of the local 'knowledges' and the wisdom of the original indigenous groups, the inter-cultural and inter-scientific dialogue and the transdisciplinarity, to guarantee the food autonomy and security, as well as greater competitiveness in the local and external markets to live well'.

The emphasis of policy on the value of local and ancestral knowledge is a matter of interest in 'closing the loop'. An insight into it may respond to the standing question of why research results are seldom reaching the small rural producer and why innovations are not taking place in this setting.

To better analyse the issue of endogenous development, it is interesting to bring the experience of the Project for Capacity and Theory Building for Universities and Research Centres in Endogenous Development (CAPTURED), under the auspices of the Ministry of Foreign Affairs of the Netherlands, in aiming to develop capacities for the research and use of local knowledge. The Project in Bolivia was conducted by AGRUCO (2008), an R&D centre of the UMSS (Cochabamba). Box 4.2 summarizes the evaluation of the Project.

4.5 THE HEALTH SECTOR INNOVATION SYSTEM

4.5.1 Overview

The development and strengthening of the health system has always been considered a priority in Bolivia. Administrative decentralization initiated in the second half of the 1990s brought important benefits to the sector, mainly because it allowed the presence of the State and, along with it, resources for areas of the country where access to health services was virtually non-existent. Thus, popular participation and administrative decentralization forged a decentralized concept of the health sector, revealing deficiencies particular to the different regions of the country, including cultural, social and economic obstacles and providing local solutions (Aramayo Carrasco, 2009).

The health innovation system is composed of a large number of public and private organizations. In the public sector, the Ministry of Health and Sports (MSD) leads the system, not only in policy definition but also in conducting research and services through its dependent institutes. Two vice ministries support the policy and implementation tasks of MSD: the Vice Ministry of Health and the Vice Ministry of Traditional Medicine and Intercultural Affairs. The government has also established, within

BOX 4.2 EXPERIENCE IN THE CREATION OF CAPACITIES FOR ENDOGENOUS DEVELOPMENT

AGRUCO was able to establish a learning community interested in subjects like inter- and intra-culturality, inter-scientific dialogue, inter-civilizational dialogue, participatory research methodologies that recognize the wisdom of indigenous peoples, decolonization, integrative logical understanding of history and culture,and transdisciplinarity. AGRUCO achieved a shift in the main concepts it is working with: from agro-ecology, biodiversity, sustainable endogenous development, reciprocity towards indigenous economies, legal and political issues like plural-nationality, autonomy, plural legal systems, communitarian socialism, wellbeing and other concepts.

In the current political reality of Bolivia it is important to take the 'Living Well' discourse and ideology to the next level and AGRUCO contributed with support to the NDP, the framework of Intercultural Governance and other policies, programmes and projects. The plural-national education system should future strengthen this process and AGRUCO provided an example that has the potential to be expanded in Bolivia and Latin America.

The evaluation identified the following challenges that AGRUCO faces:

- The process of qualification at postgraduate level, especially in the Master's, has improved since 2010 but needs a strategy to achieve a higher percentage of students' qualifications.
- The PhD programme needs co-funding in order for PhD students to complete their research projects.
- AGRUCO could increase its presence in the international scientific community by making a clear strategy in identifying partners who can contribute to the dissemination of scientific and other achievements.
- There is potential for AGRUCO to work in new areas of knowledge and topics in coordination with and complementary to other partners, such as the health sector, technology, migration and urbanism, the multinational State, religion and energy.

The evaluation concluded that AGRUCO is well positioned in the Bolivian socio-political context, has contributed to achieving major institutional impact, has developed a set of innovative education programmes at different levels and has also contributed in development efforts.

Source: Orellana Halkyer and Brouwers (2012).

MSD, a National Advisory Unit for Project Planning and Coordination. This unit has a research and technology sub-unit and sets priorities for research.

Within the MSD, the programmes that contain research activities are: National Control of Tuberculosis, Fight against Great Endemic

Diseases, National System of Health Information, Epidemiologic Shield Programme (EE), Integral Health Project and the National Institute of Health Laboratories (INLASA). It should be highlighted, however, that no programme of the Ministry carries out clinical research.

The greatest concentration of MSD, rather atomized, resources assigned to R&D goes to INLASA (5 per cent), which focuses activity in the biomedical area (almost 80 per cent); the rest of its activities are dedicated to public health. Although quite far from INLASA, EE comes in second place (2.8 per cent) in the share of resources assigned to research. The National System of Health Information had the lowest share of expenditure. The Ministry concentrates a great part of its energy and resources in epidemiological control.

A key actor of the health innovation system is the University of la Paz (UMSA), which operates six large research centres with different focuses: (1) Research Institute in Health and Development (INSAD); (2) Institute of Genetics; (3) Bolivian Institute of High Altitude Biology (IBBA); (4) Institute of Laboratories Services for Diagnosis and Health Research (SELADIS); (5) Institute of Research in Pharmacology and Biochemistry; and (6) Centre for Information and Documentation of Medicine.

The greatest concentration of R&D projects is found at IBBA, where clinical and public health research has high priority; these two research methodologies, when added together, represent 73 per cent of the total number of projects. Within the IBBA clinical research, 'non-communicable diseases and addictions' are the focus of most research. In public health research SELADIS focuses on 'nutrition and the environment'.

In the private sector, the coordinating network PROCOSI is composed of 36 NGOs, several working in isolated communities, with tasks ranging from data collection to analysing information and developing community action. PROCOSI supports capacity development in NGOs to assist communities with prioritizing and supporting their own research and has contributed to knowledge generation, developing methodologies, for example, in supporting the intercultural-communitary family health policy, participative methodologies and proposals for communication of social change.

Mc Farren (2006) has analysed the decision-making system in the area of health, pointing out that it is necessary to strengthen the capacity of community members to play an active role in the health innovation system, particularly to extract and build on existing experiences within NGOs, and determine capacity needs and strengthen the link between academia and the NGO sector.

Advances in health research have been widely documented, although much remains to be done in the way of diffusion of research results, as explained by de Pardo Ghetti (2012) in her analysis of the publication of the *Bolivian Medical Gazette*. Both at the level of public and private research organizations such advances have included the development of vaccines and better knowledge of the health–anthropology relationship when dealing with native populations (Mc Farren, 2006).

At the university level, one result of a SIDA/SAREC financed project at UMSA was introducing biogas production in small villages and new medication against parasitic diseases. In this particular project, cooperation contributed to the development of a research culture that did not exist before, in particular the use of funding mechanisms.

An interesting example of a multidisciplinary project that has yielded an extensive number of international publications is that of the Tsimane Health and Life History Project (Kaplan and Gurvey, 2013), a joint health and anthropology project aimed at understanding the impacts of ecology and evolution on the shaping of the human life course.

4.5.2 Research Policy in the Health Sector

The disconnect between health sector research and policy definition in the attempt to achieve health sector goals has been a common policy problem in countries in Latin America. Policy makers often lack proper knowledge on what works and what does not, either because relevant research does not exist or because their results are not promoted and diffused efficiently.

In an attempt to understand and measure the extent of this phenomenon in ten countries in the region, Urcullo et al. (2008) made an important survey-study to list and analyse the principal themes in terms of priorities for health policy and research in the following areas: health financing; human resources for health; and role of the non-governmental sector in health. The research used three methods of inquiry: (1) quantitative analysis of closed-end questions restricted to predefined priority listings; (2) qualitative analysis of open-end questions about country-specific policy and research priorities in each thematic area; and (3) review of the existing scientific and grey literature.

Some of the results of the survey-study provide the answers needed to analyse how near the health innovation system is in 'closing the loop' in Bolivia. Table 4.16 shows the specific questions that were asked and the responses. In the table, R represents the coherence indicator as the coefficient of determination (or R2), between two different rankings of priorities. When analysing responses to the different questions the study of Urcullo et al. shows high agreement with policy decisions in the academic sector in the country's research agenda. This result must be compared to that of the survey carried out in this project (discussed in the next section)

Table 4.16 Opinion of experts in respect to 'closing the loop' in the health sector of Bolivia

Question in the survey	Response and R2 coefficient
Does government respond to the needs of the population?	Perhaps to some extent (R2 = 0.40)
Is scientific evidence being generated to assess, change or improve current policies?	Yes, very much (R2 = 0.83)
Should scientific evidence aimed at assessing, changing or improving current policies be generated?	No (R2 = 0.28)
Is scientific evidence being generated to support policies that respond to the needs of the population?	Perhaps to some extent (R2 = 0.35)
Do respondents value scientific evidence to support policies that respond to the needs of the population?	Yes, very much (R2 = 0.85)
Is research moving in the right direction?	Difficult to say (R2 = 0.23)

Note: R2 values range from 0 to 1, where 0 represents no coherence at all and 1 represents full coherence.

Source: Urcullo et al. (2008).

showing agreement, although the results of the research itself have not impacted strongly in tackling the sector's challenges.

4.5.3 Policies and New Guidelines for Health Research

Since 2006, the government has followed the priority concerns of the past and revamped its policy vision in several ways; for example, traditional medicine has been revaluated and social policy has been defined as a high priority by the government. The government has also requested that there are specific diseases that require mandatory reporting and are part of the epidemiologic profile of the country.

Present policy includes research and technology management incorporates high technology for the diagnosis, treatment and rehabilitation according to the population needs, guaranteeing the preservation of intellectual property rights of the traditional knowledge of indigenous and originary communities. Further policy calls for the development and implementation of national technology progressively and sustainably to substitute imported technology, with the possibility of introducing into the regional and world markets alternative products and procedures.

There are regulations that apply to medical ethics, patents and medical products. INLASA has a laboratory in charge of quality control and dealing with patents is a responsibility of MSD, which is now working to make the national legislation (and statistics) compatible with the Andean Decision 486 (Common Law on Industrial Property).

The private sector has normally been absent from health research, although the national pharmaceutics industry has been growing steadily, having exported some products to neighbouring countries. The private sector, however, has not been part of the definition of research agendas and its linkages with universities have also been limited.

In March 2008, after a period of several months of preparation and under the leadership of the MSD, the institutions involved in the health sector were invited to participate in a meeting to design a National Agenda of Research Priorities in Health. After an in-depth analysis and following the recommendations of the meeting, the Agenda for Research was adopted (MSD, 2009). The MSD prioritized ten sub-agendas around which the Agenda was built by the key stakeholders; these contain priority themes and research areas and 216 lines of research to implement them. It is noted that the Agenda lacks time as a key element in the definition of actions for the adequate functioning of the innovation system. Moreover, the number of prioritized research lines is too large and no financial resources are quantified to implement the research lines.

4.5.4 The Use of Ancestral Knowledge in the Health Sector

Different governments in the past two or three decades have recognized the value of ancestral medicine and its practices. In 1984 the Bolivian Society for Traditional Medicine (SOBOMETRA) received recognition. The same year, SOBOMETRA and the Ministry of Health subscribed to an agreement that recognized the integration of 'native medicine, official medicine and alternative medicine' in the health system of Bolivia. By 1987, the norms for the exercise of traditional naturist medicine were approved. Also in 1987, the Bolivian Institute for Traditional Medicine 'Kallawaya' was created by Law. In 1996 and 1997, a new law and its regulations were approved by which natural and traditional medical products were regulated. In 2001, new norms were approved regarding the regulation in the use of traditional products, and the Strategy for Poverty Reduction and the Basic Indigenous and Originary Health Insurance was also approved in 2001.

The new Constitution of 2009 affirms the need to provide and guarantee the respect for, the use, research and practice of traditional medicine, recuperating the knowledge, values, ancestral practices and wisdom of all nations in the new pluri-national State of Bolivia.

SOBOMETRA is an NGO dedicated to the research of ancestral knowledge on the use of medicinal plants and other healing practices in the indigenous and peasant communities of Bolivia. Accompanying its research efforts, it trains and further promotes research into traditional medicine. It also supports the use of more than 300 medicinal plants. It is estimated that 80 per cent of the Bolivian population use traditional medicine developed several centuries ago.

Ancestral practices include phyto-therapy (use of plants), geo-therapy (treatment with mud), hydro-therapy (water treatment) and helio-therapy (solar baths), with the idea that these treatments help to reconcile the individual with the natural elements. There are also therapies that integrate nutrition with psychological support. Health in this vision is the result of a harmonious integration of human beings, cosmos and 'mother earth'.

4.6 SURVEY ON 'CLOSING THE LOOP'

4.6.1 The Sample

During the course of the project in Bolivia a survey was conducted in two periods in 2013 among members of the research community to test how far 'closing the loop' was actually occurring. For this purpose, 250 (bi-lingual) questionnaires were sent by both email and personal contact; a total of 75 (30.0 per cent) responses were received. This was a random sample selected from approximately 600 full-time researchers and 600 teaching staff (academics) with research records, working mostly in the five major public universities and three private universities of the country. Of the 75 responses 60 per cent are full-time researchers and 40 per cent are teaching staff with a research record.

The survey did not specifically aim at donors or policy makers. However, about 50 per cent of the academics were in decision-making positions, mostly in the private enterprise sector and fewer in the public sector. Also, many of the consulted researchers had been in policy-making positions at some time in their careers. The emphasis on the research community lies in the fact that in a developing economy, such as Bolivia, it is principally the research community that has the major responsibility to inform and influence decision making.

The survey assessed the current state of NIS in Bolivia using two methods: (1) closed questions for the overall assessment of the research and innovation system in Bolivia, and (2) open questions in order to capture those missing elements that were not contemplated in the first part.

It is considered that the result of the survey is highly representative of the qualified opinion of the research and academic community because of the level and experience of those that have responded: 26 PhDs, 33 MSc and 15 BSc (one respondent did not provide information on academic degree); of these 54 were male and 21 female, and 60 per cent were between the ages of 31 and 45 years. The respondents belong to different social and economic sectors of the country: agriculture, health, economics, technology transfer, climate change, energy, among others.

4.6.2 The Results

Relevance of research

In terms of the relevance of the research as a policy informing and influencing tool and the implementation of its outputs, researchers consider first that there is a large communication gap between the scientific community and decision makers; 87 per cent of the surveyed researchers and academics believe that such a gap exists.

It is claimed that the gap occurs as researchers are not well aware of the new legal and economic framework existing in the country, while at the same time public officials lack understanding of what technical contributions can do for development. In the former case, Figure 4.5 shows the relevance of projects towards problems identified in public policies. The figure shows that respondents consider that 16.13 per cent of projects are

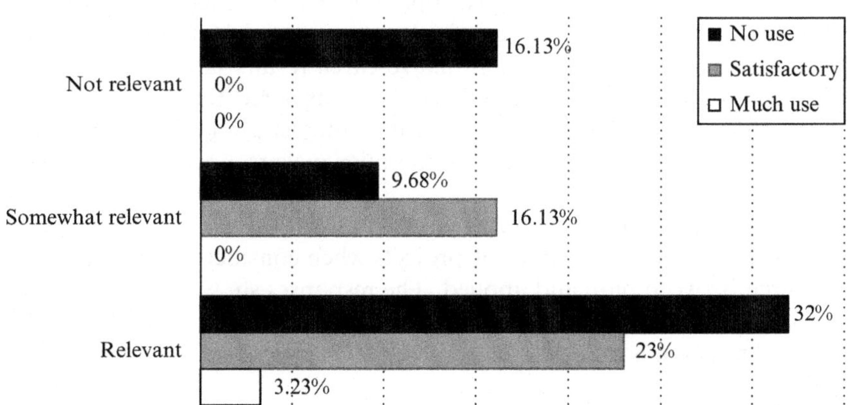

Source: Survey.

Figure 4.5 Relevance of projects to public policy and use

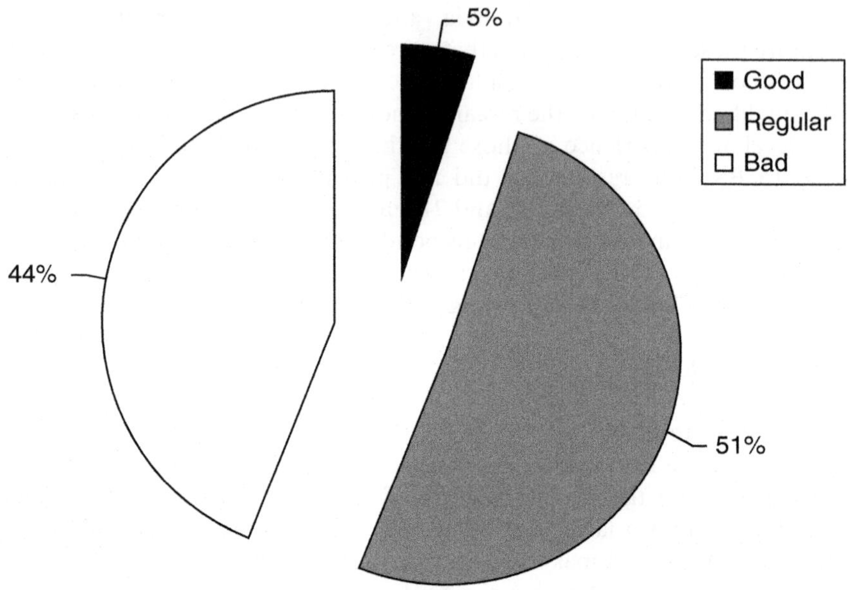

Source: Survey.

Figure 4.6 Perceived quality of research

not relevant to public decision makers and thus they do not make use of research results. The percentage of opinions that consider that projects are only somewhat relevant is larger, 25.81 per cent, and 16.13 per cent of the respondents express the opinion that research results satisfy public policy demands in a satisfactory way. It is also considered that policy relevant research is carried out in 58.23 per cent of projects, but of these only 3.23 per cent of results are of much use to policy makers, while 32 per cent are not.

From a more limited number of responses (33) it is also possible to assess the perceived relevance of projects when considering the balance of research between pure and applied. The responses show that the majority of researchers (81 per cent) considered that the balance between basic and applied research is inadequate, as the former carries too high a weight.

Quality of research

Figure 4.6 illustrates the assessment made of the quality of research that is taking place in the country. It is noted that 44 per cent of responses indicate the bad quality of research. The presence of policy makers in the

Table 4.17 Quality of research in projects where decision makers did or did not participate

Quality of research in projects where decision makers participated in projects			Quality of research in projects where decision makers did not participate			Total
High quality	Regular quality	Bad quality	High quality	Regular quality	Bad quality	
1%	4%	4%	4%	47%	40%	100%

Source: Survey.

Table 4.18 Utilization of research results

Use of research results when projects were known to decision makers			Use of research results when projects were not known to decision makers			Total
Much use	Satisfactory	No use	Much use	Satisfactory	No use	
4%	46%	50%	2%	45%	53.3%	100%

Source: Survey.

projects did not necessarily lead to an improvement in the quality or outcomes of the research. Decision makers participated in 9 per cent of the research projects conducted by the researchers surveyed in this study. Table 4.17 shows the perceived quality of research projects in which decision makers did or did not participate.

The above response is in line with a bibliometric analysis that compares the number of published scientific results in Bolivia with those in other countries of the region in an H Index measuring the impact of research from the number of citations given to a specific number of publications by a given author. Brazil leads the group of 11 countries with an H Index of 305, while Bolivia is placed in last position of the ranking in this group with an H Index of 71 (SIR, 2011). This ranking of the country is low and, as noted above, the small impact of research not only in scientific publications but also on policy and innovation can be explained in part by the fact that researchers are not publishing internationally and university academic staff concentrate more on teaching than research.

Utilization of research for policy

Table 4.18 shows the utilization of research results by policy makers in both the cases when they have known or not known of the research project.

It is noted that even though results were known to policy makers and in some cases they participated in the project themselves, the utilization rate is still very low. The result of this response can be compared to that which indicates that when decision makers did not know of the project, at the time of execution, the use of research results shows a similar pattern.

From the above results it may be concluded that, independent of relevance, only about 3 per cent of the total number of research results are highly utilized by public policy makers, while about 40 per cent are only satisfactorily used and around 58 per cent are not used at all.

It could also be concluded from this result that research is being only partially focused in those areas of public policy interests or interest of public decision makers, but still results do not inform or influence decision making.

Of the total number of researchers that responded to the survey, 39 per cent considered that they never have the opportunity to present their research results to policy makers, around 47 per cent stated that they rarely have the chance to do so and only 11 per cent of the researchers said they always have the opportunity to present their research results to policy makers.

Despite the lack of connection between projects, results and policy makers, many researchers found their project important at local or national levels (78 per cent) and contained policy-oriented research problems (52 per cent); only 22 per cent felt their project did not have an impact at the local or national level, but did contain policy-oriented research problems (62 per cent).

For those projects whose researchers claim that their results had an impact at the local or national level, 17 per cent had an impact on policy, 44 per cent on social development and 39 per cent on economic development.

In order to better understand the use of research results, it is useful to further analyse the publication of research findings by the scientific community. In 2006 a survey was conducted to understand the ways the Bolivian scientific community transmits its research results (Aguirre-Bastos et al., 2007). The survey was answered by 171 researchers in public and private organizations, of which 53 were public universities and 55 private universities. Of the respondents, 140 were male and 31 female. In the sample, 22 had a PhD degree, 94 an MSc, 55 a BSc. The result of the responses to the questions on diffusing research results in the survey is given in Table 4.19.

The result here shows that oral means continues to be the preferred way of communication. This fact is of relevance to 'closing the loop' as

Table 4.19 Means of diffusing research results by the scientific community of Bolivia

Means of diffusion	Number of responses
Papers published in local periodic journals	50
Papers published in international journals	27
Papers/reports published in occasionally published local journals	36
Papers/reports published in occasionally published international journals	22
Newspaper articles	27
Papers and reports published online	43
Occasional interviews on TV and radio	32
Periodic interviews on TV and radio	10
Presentations in congresses, seminars and equivalent	88
Do not publish	39
Others	5

Note: The respondents were free to identify more than one means.

Source: Aguirre-Bastos et al. (2007).

research results provided by this means are normally very short-lived, as decision makers or other users of research results will seldom use them, unless they are backed by written communication.

Funding

A key element influencing the performance of the STI system is of course financing. As already discussed, government and private expenditures in R&D and innovation in Bolivia are very low; the survey showed that 91 per cent of researchers consider that the financing system does not function adequately, impacting on the rather poor quality of projects. Such a situation constitutes a major lack of incentive for research and is responsible for the large number of academics that do not dedicate full time to research or strive to reach better quality research outcomes.

Table 4.20 shows the different factors that are considered in the formulation of research projects. As can be seen, compliance with national development goals is not as high as would be expected when research outputs are supposed to inform and influence public policy makers.

Table 4.20 Important factors at the time of formulating research projects (multiple responses)

Question	Response	Responses	
		Number	Percentage
Which of the following factors do you consider the most important when you formulate your research projects?	1. Compliance with national developmental goals	28	28.9%
	2. Compliance with faculty/ department research priorities	12	12.4%
	3. Compliance with priorities of funding agencies	15	15.5%
	4. Academic impact	22	22.7%
	5. Other	20	20.6%
	Total	97	100.0%

Source: Survey.

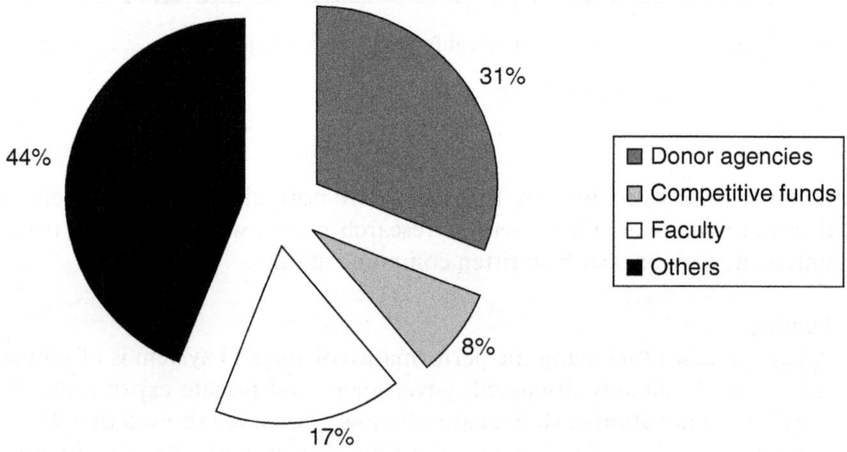

Source: Survey.

Figure 4.7 Sources of funding

Figure 4.7 shows the percentage distribution of the sources of funding for the projects. The result is consistent with the indicators produced by the Vice Ministry of S&T as discussed in the description of the STI system. The preference for donor funding is twofold; first, funds are available, even

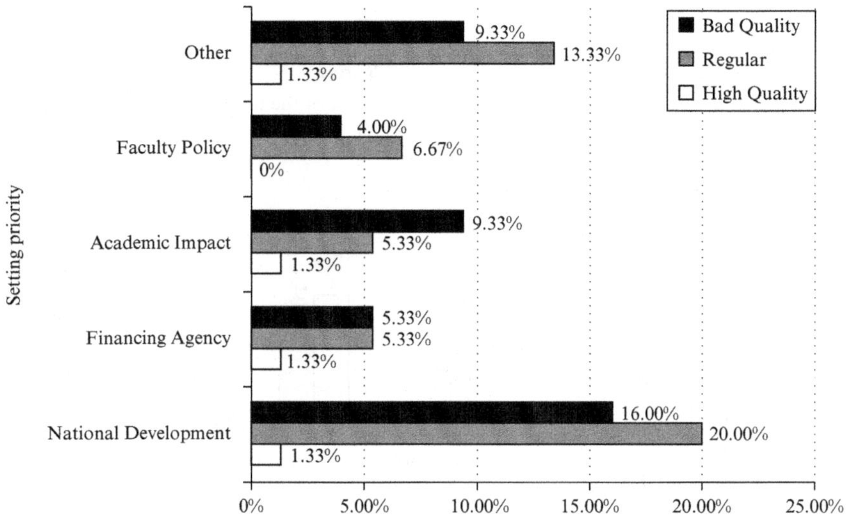

Source: Survey.

Figure 4.8 Priority setting and quality of projects

if in limited amounts, and second, they are accessible without too much bureaucratic processes involved. On the other hand, researchers prefer donor funds as they are not attached to government policies (and politics), although research results from donor funded project might not be readily utilized by public decision makers. This preference to donor funding comes both from young and also experienced researchers.

Figure 4.8 shows the distribution of projects according to who sets priorities. When projects are pursued following priorities set by national development objectives, it is noteworthy that they are perceived as of average or bad quality. This is a hindrance to policy makers at the time of making decisions and leads to a phenomenon whereby the use of outputs of bad quality might replace high quality. On the other hand, a large number of researchers consider that the measures taken by funding agencies to align research objectives with national development goals are insufficient and these agencies are not functioning adequately.

In spite of the preference for donor funding, it is the form of funding that allows the least academic freedom for project design, as shown in Figure 4.9. This may be explained by the existence of 'country strategy' definitions of donors that fix guidelines for what can be done in their view.

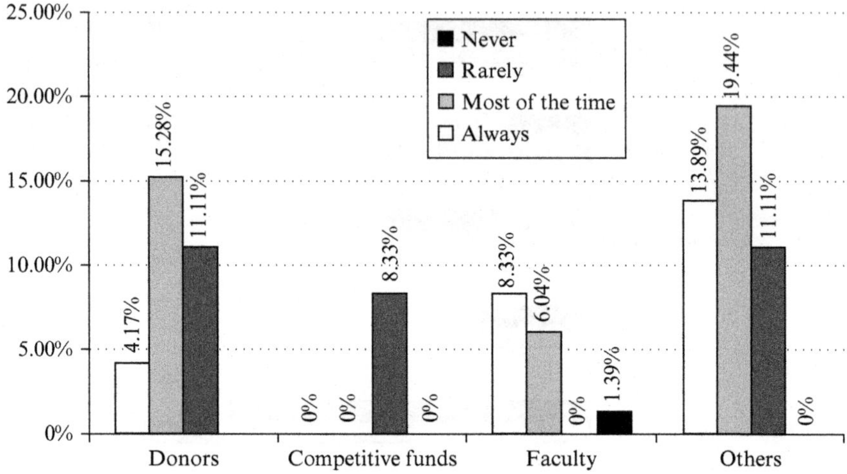

Source: Survey.

Figure 4.9 Financing and freedom of research

4.7 DISCUSSION AND CONCLUSIONS

4.7.1 'Closing the Loop'

'Closing the loop' in Bolivia can be broadly analysed from two perspectives. The first by the utilization of research outputs for policy making and the second by their use for social and economic innovations, which in turn may contain policy issues as well.

In the first case, the use of research outputs for policy is weak for different reasons as identified from the survey among the scientific community and some sector studies. Social research on STI is quite limited and needs to be enhanced. Unfortunately, the present National Plan for STI does not contain research in the social sciences as a component.

Further, the channels of diffusion used by researchers towards policy makers are not the most adequate, and thus the research and academic community is not empowered enough to trigger changes for the use of evidence based decisions.

For the country to benefit from research outputs for policy making there needs to be interest not only in the scientific communities but also the policy and political establishment that sets priorities and provides incentives to the NIS. As in the past, policy makers react more to political or social pressures than when faced with scientific evidence. Thus, the academic community is called upon to produce a change of culture in the way decisions are made in the country.

The use of research outputs for economic and social innovations is more pronounced but still insufficient to trigger in-depth social and economic changes. There are observed examples that show that 'closing the loop' is possible, as there exists, even if limited in number, the human potential for obtaining quality research outputs. To enhance the process, there is a set of conditions that must necessarily be met, not least the formulation of adequate policies.

One of the key elements for applying research outputs to social and economic demands is funding. Public (and private) expenditures in research and innovation, and for higher education, must be substantially increased and efficient financial mechanisms established.

When resources are found, and that is not always the case, these are too small to actually produce an impact. Today, researchers must find resources from any available source, and in this case research projects are those prioritized by private or bi- or multi-lateral cooperation funding agencies, not necessarily designed to respond to a policy demand.

The above means not just adequate but also sustained and transparent funding, the presence of professional research managers, ethical standards and accountability in the use of public funds. In general, for the NIS to operate efficiently, a culture of evaluation needs to be introduced at all levels, so that research agents will be backed by a scientifically conducted process of measuring outputs and impacts.

In the more specific case of universities, it is satisfactory to find that many university projects deal with subjects of direct development relevance. But often these research interests have not been fully implemented in the form of extensive operation with research users, such as industries and authorities, as well as other parts of society, especially poor and indigenous communities. Certainly, it is necessary that university research environments should be more competitive and globalized, a culture of innovation must be generated, new career opportunities must be created, the best talent should be attracted and a creative and open but rigorous, demanding and empowering environment should be generated.

Rapid technological change, globalization and other international developments make it necessary for universities to upgrade their activities from those of more or less advanced teaching institutions to modern universities. Bolivian universities must become able to support national development through research based solutions to development problems and, in particular, production of university graduates with a relevant research background. A renewed higher education system destined to attend the development of positive attitudes fostering creativity, entrepreneurship and technology and non-technology based innovations is absolutely necessary to face future challenges.

From the sector analyses, it is found that agriculture has always been given a high priority and consequently, agricultural research has been on different government's agendas for several years. Closing the loop has been possible when dealing on a product-by-product basis. One evident example is that of the Andean grain quinoa, which is today successfully commercialized nationally and internationally, after scientific research in the 1960s and 1970s provided the basis for improving commercial varieties.

Considering this example, it is clear that selective interventions should be sought focusing on strategic areas, those that can help the country to accumulate the know-how and capabilities for pursuing more sophisticated production and commercialization methods. It has been suggested by the 2012 Global Conference on Agricultural Research for Development that the policies that will improve the performance of the sector innovation system should look at foresight, capacity development and partnerships. These are not necessarily a priority in the present policy.

In considering 'closing the loop' in the agricultural sector innovation system, it must also be understood that the context for agriculture is changing rapidly and radically, and among the challenges to be met is that markets, not production, increasingly drive agricultural development. This situation contrasts with present government priorities for this sector. In particular, the strongly centred, State-managed research and innovation agenda should consider the enterprise as the central player in the innovation system and the need to follow more traditional or innovative entrepreneurial approaches.

4.7.2 Policy to Strengthen the National Innovation System and Improve the Impact of its Research Outputs

Science, research, technology development and innovation are today the main drivers of economic growth and social improvement. The country cannot be expected to face its present and future challenges of sustained and inclusive development and competitiveness without them.

The complexity of the process involved in knowledge generation and use requires a system's approach. Bolivia has already considered such an approach in the past and it is present in today's NDP. However, in practice STI policies are still defined in a linear mode from a strong supply-side perspective. Weakness of the system limits further impact, especially when transferring research results for social or economic innovations.

Today, policy and research agendas are being set by different agents of the innovation system. On the one hand, the Vice Ministry of S&T has adopted a National STI Plan in consultation with, principally, the research community. On the other hand, sectors such as the universities and other government agencies are defining their own agenda, causing a disconnect in policy for research and innovation.

Further, there is no analysis on the impact of implicit policies affecting the NIS. Unless there is a better understanding on how such policies affect the system and what could be done to better complement explicit with implicit policies, no policy or research agenda will be strong enough to face development challenges and may be subject to permanent changes depending on the will and understanding of individual policy makers. The need for a more coordinated policy framework is certainly one way to close the loop.

A clear understanding of the innovation process is crucial for policy definitions. From the analyses of existing documents, it is clear, first, that there is a persisting difficulty to understand the concept and practice of innovation by all agents of the NIS, and further, that the present understanding does not accommodate the new existing paradigm of open innovation.

The existing understanding of innovation and the definition of the NIS (as well as the conceptual approach to STI development) miss a key element in the production of innovations, which is the recognition of the enterprise as a key actor of the process. Such lack of recognition does not create an adequate business context for enterprises to lead innovation processes. On the other hand, policies expect that the community at large, social cooperatives and similar organizations as well as universities will be responsible for the development of innovations. While it is possible that these organizations can in fact develop technology and innovate, it is not clear how they are expected to do so, without the proper funding, human resources and linkages with a research base.

Bolivia has adopted an STI policy and strategy that relies strongly on the potential capacities of traditional knowledge and local wisdom and their combination with 'western knowledge'. This approach relies on a set of assumptions stemming from a diagnosis that is strongly ideologically oriented and does not recognize the full extent of both the strengths and weaknesses of previous STI policies, nor the intensive learning processes that have taken place in the country, the region and worldwide.

In this diagnosis it is correctly stated that scarce financial support for research and the dependence on foreign technology hampered development. However, it is quite evident, contrary to the diagnosis, that past policies were aimed at developing local potential; in fact, this has been the guiding vision in Bolivia, as well as in Latin America at large, and was developed by the Latin American School (Martínez Vidal and Marí, 2002). Further, past policies were addressed to strengthen the structural functions of the NIS.

In Bolivia, as in many other developing countries, government rhetoric on the importance of STI in development should be matched by effective policy implementation. STI policies are only effective when government can formulate and implement them effectively, for which is needed a more secure politico-institutional foundation and the capacity to enforce regulations. Thus, it is clear that the main reason for the difficulty in building up an adequate capacity in STI in Bolivia was and still is insufficient government capabilities and effectiveness.

NOTES

1. www.boliviaentusmanos.com (accessed 20 January 2015).
2. For a discussion on the emergence of the system's approach in Bolivia, see Chapter 2 in this volume.
3. The term is used in the plural following the Plan, signifying the existence of many streams of knowledge, each coming from the different existing cultural identities in the country.
4. A more detailed description of the contribution of Bolivian universities to social innovation and inclusive development will be provided in a forthcoming book to be published by UNIDev Net.
5. Estimate by the authors from various data sources.
6. www.fdta-valles.org.
7. www.fundacionpatino.org.

REFERENCES

AGRUCO (2008), 'Aprendiendo el desarrollo endógeno: Construyendo la diversidad bio-cultural', AGRUCO, COMPAS, Plural editores, La Paz, May, available at http://www.compasnet.org.

Aguirre-Bastos, C. (1997), 'Las Pequeñas y Medianas Empresas en el Sistema Nacional de Innovación', *Academia Nacional de Ciencias de Bolivia y Centro Internacional de Investigaciones para el Desarrollo, Canadá*, Publicación de la Academia Nacional de Ciencias, La Paz.

Aguirre-Bastos, C. (2001), 'El desarrollo del Marco Institucional y de Políticas de la Ciencia y la Tecnología en Bolivia', in *Contribuciones a la Historia de la Ciencia en Bolivia*, Academia Nacional de Ciencias e Instituto Panamericano de Geografía e Historia, No. 519, México, DF, pp. 3–12.

Aguirre-Bastos, C. (2002), 'Lineamientos para una Política Pública para el Desarrollo de la Ciencia, la Tecnología y la Innovación en Bolivia', Document prepared for the Bolivian System of Productivity and Competitiveness of the Ministry of Economic Development, La Paz, December.

Aguirre-Bastos, C., F.A. Mercado, J.L. Aguiree et al. (2007), *Convergent*

Technologies in the Andean Countries: What is Being Done About Them in the Andean Countries?, Final Project Report to IDRC, Catholic University of Bolivia, La Paz.

Aguirre-Bastos, C., A. Hall and J. Jiggins (2010), *Report of the External Review of the Innovation, Technology and Society (ITS) Program at the International Development Research Centre of Canada*, Ottawa, 31 July.

Alänge, S., S. Scheinberg, C. Arribasplata and R. Jimenez (2004), 'Cluster analysis', Results of a survey/study presented at the Innovation Cluster Workshop, La Paz, 30 July.

Antonio, K. (2011), 'The challenges of developing a sustainable agro-industry in Bolivia: the quinoa market', MSc Thesis, Duke University, NC.

Aramayo Carrasco, A. (2009), 'Monitoring Financial Flows for Health Research 2009', Global Forum for Health Research, Geneva, available at http://www.isn.ethz.ch.

Arauco, E. and E. Gallardo Paz (2011), *La Educación Superior en Bolivia,in Educación Superior en América Latina: Informe 2011*, October, ed. J.J. Brunner and R.F. Hurtado, Santiago: CINDA, available at http://www.cinda.cl.

Carden, F. (2009), *Knowledge to Policy: Making the Most of Development Research*, Ottawa: IDRC/Sage Publications, available at http://www.idrc.ca.

Cardozo Gonzales, A. (2001), 'Evolución de la investigación agropecuaria en Bolivia', in *Contribuciones a la Historia de la Ciencia en Bolivia*, Academia Nacional de Ciencias e Instituto Panamericano de Geografía e Historia, No. 519, México, DF, pp. 159–70.

Carvajal, R. (2009), 'Fortaleciendo las Instituciones de Innovación: El caso del sistema boliviano de innovación', presentación en el Seminario Regional, Monterrey, México.

Carvajal, R. and M. Albarracín (2007), 'Sistema Boliviano de Innovación', Documento del Ministerio de Planificación para el Desarrollo, La Paz, February.

CEUB (2012), 'Estrategia Universitaria Nacional de Ciencia, Tecnología e Innovación (2012–2015)', Comité Ejecutivo de la Universidad Boliviana.

Cozzens, S. and J. Sutz (2012), 'Innovation in informal settings: a research agenda', mimeo, 28 July.

de Pardo Ghetti, E. (2012), 'Obstáculos al desarrollo de investigación y publicación en salud en Bolivia: el rol de la Gaceta Médica Boliviana', *Gaceta Médica Boliviana*, **35**(2), 53–4.

Delgado Burgoa, F.J.M. and C.G. Escobar Vásquez (2009), 'Innovación Tecnológica, Soberanía y Seguridad Alimentaria', AGRUCO/Plural Editors, La Paz, February, available at http://www.plural.bo.

Evia, J.L., M.N. Pacheco and S.R. Quispe (2010), 'Sector Informal y Políticas Públicas en América Latina: Bolivia', Programa SOPLA, Fundación Konrad Adenauer, Rio de Janeiro.

FAUTAPO (2011), 'La Quinua Real en le Altiplano Sur de Bolivia', Documento Técnico para la Denominación de Origen, Fundación FAUTAPO para la Educación para el Desarrollo, Environmental Support Program Certificate.

Flores, A. (2003), 'El Proceso de la Revolución Verde y su Repercusión en el caso de las Comunidades Campesinas de Capaña y Laguna Pampa, Provincia Linares', Departamento Potosí, MSc Thesis, UMSS.

Fundación Valles (2010), 'Memoria 2010: Innovando el campo, mejorando vidas', Informe Anual 2010, Cochabamba, available at http://www.fdta-valles.org.

Hailu, D. and R. Tsukada (2011), 'Achieving the Millennium Development Goals:

a measure of progress', Working Paper No. 78, International Policy Centre for Inclusive Growth, February, available at http://www.ipc-undp.org.

Hartwich, F. and H.-G. Jansen (2007), 'The governmental role in the process of agriculture innovation: the experience of Bolivia', Research Resumes No. 8, IFPRI, Washington, DC, available at http://www.ifpri.org.

Kaplan, H. and M. Gurver (2013), 'The Tsimane Health and Life History Project', available at http://www.unm.edu/~tsimane/results.html#pubs.

Lemarchand, G.A. (2012), 'The long-term dynamics of co-authorship scientific networks: Iberoamerican countries (1973–2010)', *Research Policy*, **41**, 291–305.

Martínez Vidal, C. and M. Marí (2002), 'La escuela Latinoamericana de Pensamiento en Ciencia, Tecnología y Desarrollo: Nota de un proyecto de investigación', *Ciencia, Tecnología y Sociedad, Estudios*, **4**, September–December.

McFarren, W. (2006), 'Bolivia – the contribution of NGOs and communities to health research', in *Supporting Health Research Systems in Latin America: Results of Latin America Regional Think Tank*, Council on Health Research for Development (COHRED), Antigua, August, available at http://www.cohred.org.

Ministry of Economy and Finance (2012), 'Economía Plural', *Boletin Informativo*, 65, 24 September, available at http://www.economiayfinanzas.gob.bo (accessed June 2016).

Moraes, M. (2009), 'Difusión de la ciencia, indicadores e indexación de revistas científicas', *Propiedad Intelectual en la Investigación y Publicaciones Científicas*, UMSA, La Paz Segunda versión, May, 35–48.

MPD (2006), 'Plan Nacional de Desarrollo del Estado Plurinacional de Bolivia: Lineamientos Estratégicos 2006–2011', Ministerio de Planificación para el Desarrollo, La Paz.

MSD (2009), 'Agenda Nacional de Prioridades de Investigación en Salud', Ministerio de Salud y Deportes, La Paz.

OIM/SELA (2009), 'La emigración de recursos humanos calificados desde países de América Latina y el Caribe: Tendencias contemporáneas y perspectivas', Organization for Migration and Latin America and Caribbean Economic System, Caracas, June.

Orellana Halkyer, R, and J. Brouwers (2012), *CAPTURED: Bolivia Country Evaluation: Report*, available at http://www.captured.org.

Orozco Ramírez, S., Á.G. Linera and P. Stefanini (2006), 'No somos juguete de nadie: Análisis de la relación de movimientos sociales, recursos naturales, estado y descentralización, NCCR North–South', AGRUCO, Plural editors.

PIEB (2011), 'Factores que incidieron en el escaso desarrollo de Bolivia, Temas de Debate', *Programa de Investigación Estratégica de Bolivia*, ed. J.A. Morales, **18**(7), December.

Schilling-Vacaflor Del Carpio, A. (2014), *Bolivia's New Constitution: Changes and Challenges, Atomium Culture*, Brussels: European Institute for Science, Media and Democracy.

SCImago (2015), *SJR–SCImago Journal & Country Rank*, available at http://www.scimagojr.com (accessed 7 May 2015).

SIR (2011), *SCIMAGO Institutions Ranking (Ibero-America)*, available at http://www.scimagolab.com and http://www.scimagoir.com.

Sunkel, O. and R. Infante (eds) (2009), 'Hacia un Desarrollo Inclusivo: El caso de Chile', Comisión Económica para América Latina, Fundación Chile y Organización Internacional del Trabajo, Santiago.

Tellería, J.L. (ed.) (1993), 'Levantamiento del Potencial Científico y Tecnológico de Bolivia', Consejo Nacional de Ciencia y Tecnología.
Thulstrup, E.W., M. Muñoz and J.-J. Decoster (2006), 'Building research capacity in Bolivian universities, evaluation of SAREC/SIDA financed project in Bolivia', Swedish Agency for Research Cooperation/Swedish International development Agency, January.
UNDP (2011), *Informe Nacional sobre Desarrollo Humano en Bolivia: Los cambios detrás del cambio; desigualdades y movilidad social en Bolivia*, La Paz: UNDP, available at http://idh.pnud.bo.
UNDP (2013), *Human Development Report 2013*, New York: UNDP.
UNDP (2014a), 'Perfil de estratos sociales en América Latina: pobres, vulnerables y clases medias', Programa de Naciones Unidas para el Desarrollo, New York.
UNDP (2014b), *Human Development Report: Sustaining Human Progress: Reducing Vulnerabilities and Building Resilience*, New York: UNDP.
Urcullo, G., R. Muñoz and R. Bitrán (2008), *Identification of Priority Research Questions in LAC Within the Areas of Health Financing, Human Resources for Health and the Role of the Non-state Sector*, Final report, Bitrán & Asociados, Santiago, February.
Velazco Reckling, E. (2002), 'Valoración Sistémica de la Problemática de Productividad y Competitividad en Bolivia', Informe al Sistema Boliviano de Productividad y Competitividad, La Paz, October.
VMCT (2012), 'Memoria de las Redes Nacionales de Investigación en Ciencia y Tecnología', La Paz, 11 May.
VMCT (2013), 'Plan Nacional de Ciencia, Tecnología e Innovación', June, available at http://www.cienciaytecnologia.gob.bo.
WEF (2014), *Global Competitiveness Index 2014–2015*, World Economic Forum, Geneva.

5. The national innovation system in Vietnam and its relevance for development

Tran Ngoc Ca

5.1 THE NATIONAL SOCIAL AND ECONOMIC CONTEXT

After 20 years of radical socio-economic reforms and international economic integration, Vietnam has achieved some significant results. The gross domestic product (GDP) growth rate has been quite high and stable at around 6–8 per cent for the last ten years. The country has increased GDP per capita from USD 440 in 2002 to around USD 700 in 2006 and reached a threshold of more than USD 1000 in 2009 and 1910 in 2013 (World Bank, 2015). GDP per capita in purchasing power parity (PPP) increased more than threefold during 1989–2013, from USD 1097 to USD 3620 in 2013. The inflation rate has been reined in to under 10 per cent after a high of over 20 per cent in 2008.

In parallel with economic growth, the Vietnamese government has put significant effort into poverty reduction. This has resulted in an incredible outcome with the dramatic drop in the poverty rate from 70 per cent in 1986 to 7.8 per cent in 2013. The unemployment rate was at about 6 per cent on average during 1996–2005, decreasing to 2.6 in 2013 (Table 5.1). Overall, Vietnam's poverty alleviation target was completed ten years earlier than the Millennium Development Goals (MDGs) promoted by the United Nations in 2015 (discussed in Chapter 2).

5.1.1 Structure of the Economy

The economic structure has shifted from being predominantly agricultural to industrial based. The recent growth of the Vietnamese GDP is partly the result of economic restructuring. In the 1990s, the Vietnamese economy mainly depended on the agricultural sector. But the share of the agricultural sector of GDP decreased gradually from 38.7 per cent in 1990 to 22.1 per cent in 2008 and 19 per cent in 2013. The contribution of the industry

Table 5.1 Economic performance of Vietnam, 2000–13

Year	2000	2002	2004	2006	2008	2010	2013
GDP growth rate (%)	6.79	7.08	7.79	8.23	6.31	6.8	5.4
Inflation (%)	−0.6	4.0	9.5	6.6	24.4	8.9	6.6
GDP (USD)	402	441	561	730	1052	1240	1910
Unemployment rate (%)	6.42	6.01	5.6	4.82	4.65	2.6	2.6
Poverty Index (%)	–	28.9	24.1	–	–	12.6	7.8
Human Development Index (HDI)	0.70	0.69	0.71	0.72	0.73	0.63	0.6

Source: General Statistics Office of Vietnam, World Bank and United Nations.

and construction sector to the national economy increased from 22.6 per cent in 1990 to 38.1 per cent in 2013 (Table 5.2).

There has not been much change in the proportion of the service sector in the Vietnamese economy, which accounts for around 38–42 per cent of GDP over the last 20 years. Although the structure of the Vietnamese service sector has become more sophisticated, other services such as science research, design and market research are less developed. Some important services like finance, telecommunications and infrastructure remain less competitive internationally.

5.1.2 Export Activities

Vietnam has pursued a trade-led growth strategy resulting in a steady increase in both exports and imports. In 2012, the value of exports reached USD 114.6 billion of which the main exports included textiles and garments (USD 15.1 billion), cell phones and accessories (USD 12.7 billion), footwear (USD 4.7 billion), seafood products (USD 6.1 billion) and agriculture products (USD 9.9 billion). The export value of rice also reached USD 3.7 billion in 2013 (World Bank, 2013).

The export value of crude oil decreased considerably in the period 2006–10 because of an increase in crude oil delivery for Dung Quat oil refinery. The export value of crude oil in 2012 accounted for only 7.2 per cent of the total Vietnamese export value (USD 8.2 billion), a significant decrease compared to 21 per cent in the period 2001–05.

Table 5.2 Trend of economic structural change in Vietnam 1990–2010

	1990	1992	1994	1996	1998	2000	2002	2004	2006	2008	2010	2013
Share of GDP (%)	100	100	100	100	100	100	100	100	100	100	100	100
Agriculture, forestry & fishery	38.7	33.9	27.4	27.7	25.7	24.5	23.0	21.8	20.4	22.1	20.5	19.0
Industry & construction	22.6	27.2	28.8	29.7	32.4	36.7	38.4	40.2	41.5	39.7	41.1	38.1
Service	38.5	38.8	43.7	42.5	41.7	38.7	38.4	37.9	38.0	38.1	38.4	42.9

Source: General Statistics Office of Vietnam and World Bank, various years.

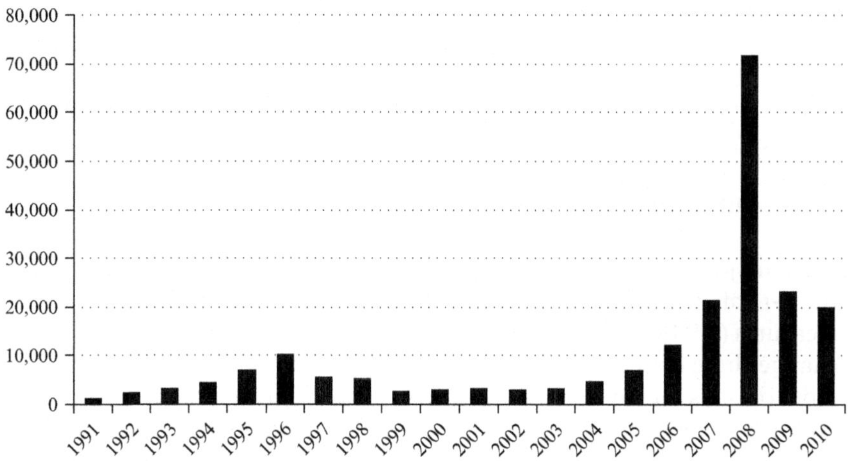

Source: General Statistics Office of Vietnam.

Figure 5.1 Vietnamese registered FDI 1990–2010 (million USD)

5.1.3 Performance of the Foreign Direct Investment Sector

Among the factors leading to socio-economic development, foreign direct investment (FDI) plays a very important role, accounting in 2008, for instance, for 31 per cent of total investment capital in Vietnam. State and non-state investment accounted for 29 per cent and 40 per cent, respectively.

Since the approval of the Law on Foreign Investment in December 1987, Vietnam has achieved an impressive performance in attracting FDI flows. FDI reached a peak with 833 projects and USD 10.201 billion in 2006, the highest numbers since 1987. After 2007, as soon as Vietnam became an official member of the World Trade Organization (WTO), the total registered capital reached over USD 20 billion, almost double that of 2006. FDI in 2008 dramatically increased after Vietnam became a WTO member, reaching USD 71.73 billion (Figure 5.1). After this time the numbers have gradually decreased to reach USD 20 billion in 2014.

The remarkable increase of FDI has played an important role in fostering production and business as well as contributing to the growth of the economy.

Contribution to GDP growth

The share of the foreign-invested sector (FIS) in GDP has increased rapidly and this sector has proved to be an integral part of Vietnam's economy. In

2008, FIS contributed approximately one-fifth of GDP. Moreover, FIS has contributed to the structural change towards industrialization. In recent years, FIS has been dominant in highly value-added industries, such as crude oil, computers, telecommunications, electronic products, spare parts and steel.

5.1.4 Other Macroeconomic Issues

Vietnam joined the WTO in 2007 with several important commitments like trade-related intellectual property rights (TRIPS), trade-related investment measures (TRIMs), technical barriers to trade (TBT) and so on. Economic changes have connected the country to the global economy and opened up more options for enterprises in finding solutions to their business problems and helps them connect with the international production networks. The WTO accession is driving further liberalization of the economy as well as its global integration. The growth of the private sector has also risen dramatically. In 2011, there were 602,300 registered small and medium-sized enterprises (SMEs) (Pham Thi Thu Hang, 2011) and the private sector accounted for some 40 per cent of GDP (50 per cent in 2005), 40 per cent of total investment (27 per cent in 2005) and 77 per cent of employment (90 per cent in 2005) (ADB, 2005; BIA, 2011).

Despite the achievements, the challenges faced by Vietnam in sustaining high growth rates through enhanced productivity and competitiveness are great. Although the poverty rate has been reduced, the gap between rich and poor has increased. International integration of the economy has brought benefits, but at the same time, some problems. The biggest challenge lies in increasing the level of productivity: Vietnam has relied primarily on investment and labour force expansion rather than different forms of innovation, which have a significant impact on productivity growth. These conditions have resulted in a low level of Total Factor Productivity (TFP) contribution in the economy, which underlines the weakness of innovation inputs. Economic growth appears to rely mainly on capital and labour, without appropriate attention paid to boost TFP. Despite significant progress, the contribution of knowledge to economic growth is still limited. During 1990–2000, TFP, including knowledge/technology, contributed only 44 per cent of GDP growth. This rate was even lower at 26 per cent during 2001–08. This is one of lowest among the Association of Southeast Asian Nations (ASEAN) (Table 5.3)

In the agricultural sector it is clear that Vietnam's success stems more from policy reforms that have taken place over the past two decades rather than the introduction of new technology (for example, seeds, equipment, practices and so on). TFP growth in Vietnam is low compared with the rest

Table 5.3 Factors contributing to GDP growth in Vietnam and other Asian countries in 1990–2008

Countries	1990–2000				2000–08			
	GDP growth (%)	Growth factors			GDP growth	Growth factors		
		Capital	Labour	TFP		Capital	Labour	TFP
Vietnam	7.3	2.5	1.6	3.2	7.3	3.9	1.4	1.9
China	9.9	3.6	0.7	5.5	9.7	4.1	0.6	5
Cambodia	7.3	2.8	2.5	2	9	4.2	3.5	1.3
Indonesia	4.1	2.5	1.1	0.5	5.1	1.4	1.1	2.5
Malaysia	6.9	3.7	2.1	1.1	5.4	1.6	1.1	2.7
Philippines	3	1.3	1.4	0.3	4.7	1	1.9	1.8
Thailand	4.4	2.7	0.3	1.4	4.7	0.8	1.4	2.5
Contributing proportion								
Vietnam	100%	34%	22%	44%	100%	53%	19%	26%
China	100%	36%	7%	56%	100%	42%	6%	52%
Cambodia	100%	38%	34%	27%	100%	47%	39%	14%
Indonesia	100%	61%	27%	12%	100%	27%	22%	49%
Malaysia	100%	54%	30%	16%	100%	30%	20%	50%
Philippines	100%	43%	47%	10%	100%	21%	40%	38%
Thailand	100%	61%	7%	32%	100%	17%	30%	53%

Source: CIEM (2010).

of the region and the world. Moreover, there are large differences in productivity in Vietnam between formal enterprises within the same industry and between formal and informal sectors. Lack of resources for investing in technology innovation tends to prevent private and informal sectors from becoming more innovative.

5.2 NATIONAL DEVELOPMENT PLAN

Vietnam embarked on a path towards radical social and economic changes at the beginning of the 1980s, with various types of reforms and experiments. But only by the end of the 1980s did real reforms take place in all social and economic aspects starting with the Party Congress in 1986. The reforms were reinforced when Vietnam faced severe difficulties at the time of the collapse of the Soviet economic bloc in the 1990s. There was a great need to open up the economy and generate greater diversification of international relations that would lead to more trade and foreign inward investment. With regards to science, technology and innovation, the policy landscape experienced several periods of change. Early reforms came during the 1980s, with the autonomy of research and development (R&D) institutions, when direct contracts with industry were allowed as well as the ability to operate like an enterprise. Flexibility of R&D organizations was increased to develop and provide a full range of services such as technology transfer, consulting services, experimental and pilot manufacturing. Financing has been diversified, while these organizations can retain profits and have legal authority to seek bank credit. There began the process of privatization of R&D activities accompanied by improvements in the legal framework on intellectual property rights (IPR).

More recent changes occurred with a number of important milestones and documents such as: the science and technology (S&T) Strategy to 2010 in 2003; projects to develop the technology market (2005); international integration in S&T (2007); Law on S&T (2001); Law on IPR (2005); Law on Technology Transfer (2006); Decree No. 115 on the autonomy of public R&D organizations (2007); Decree No. 119 on firms' support programme (2003); Decree No. 80 on S&T enterprises; creation of funds for S&T development (2007); the State Agency for Technology Innovation (2007); and High Tech Law (2008).

The evolution of Vietnam's institutional framework for S&T development is depicted in Figure 5.2.

5.2.1 Strategies and Priorities for Science and Technology Development

The S&T Strategy for Vietnam to 2010 provided several directions and priorities for S&T development. More recently, in 2012, the prime minister issued a Decision on a new Strategy for S&T development to 2020. This Strategy has the objectives: to develop in a synchronized manner social sciences and humanity, natural sciences, technical and technological science; to make S&T become a key motivating force and meet the basic requirements of a modern industrial country. By 2020, it is planned that a number of Vietnam's S&T fields will have reached the advanced, modern level of the ASEAN region and the world. Priority areas are information and communications technology; biotechnology; new materials technology; machinery and automation technology; environmental technology; energy S&T; construction S&T; marine S&T; management S&T; space S&T. The Law of S&T promulgated in 2001 was revised and approved in 2013.

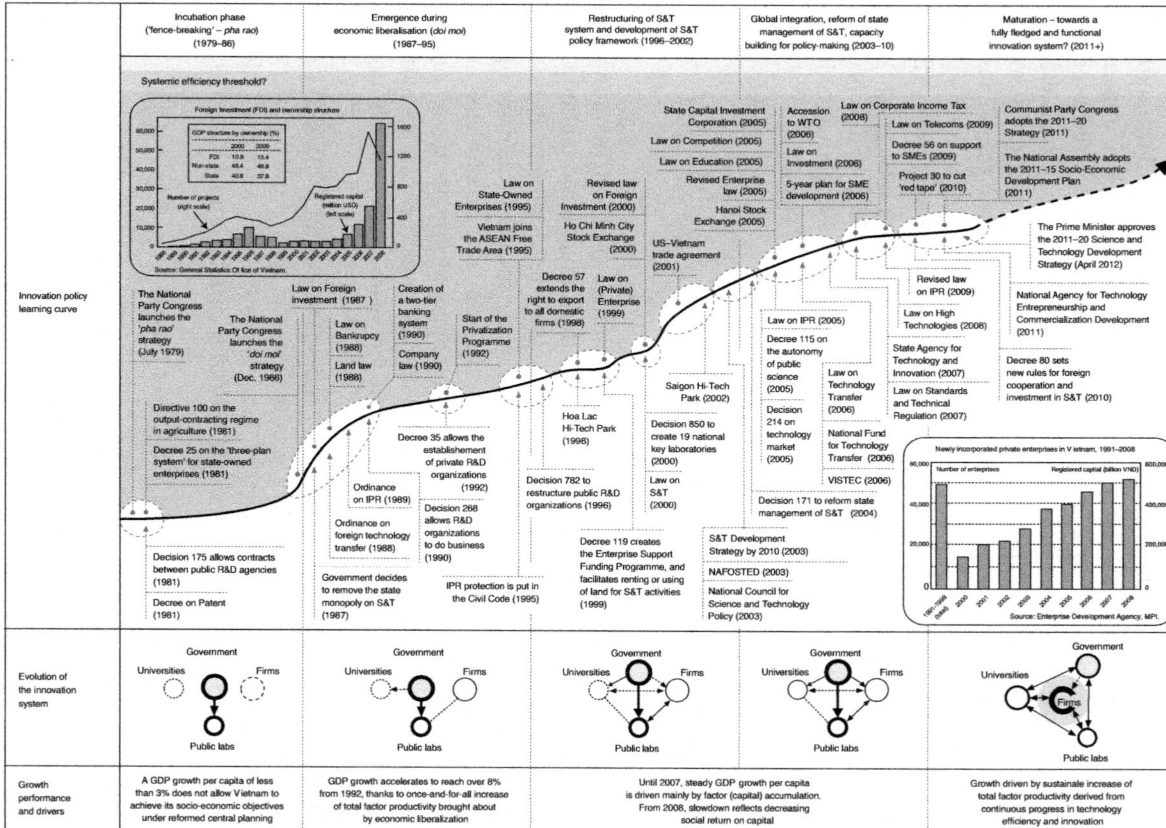

Source: OECD/World Bank (2014).

Figure 5.2 Evolution of the institutional framework for S&T in Vietnam

5.2.2 The Government S&T Programmes and Techno-economic Programmes

From the 1980s, following the five-year plan, Vietnam organized the state-level S&T programmes in key fields to serve socio-economic development (in recent years, about 15 to 20 programmes have been organized every year). Half of these programmes relate directly to the creation of new techniques and technology for application in production. The ministries and provinces also have S&T programmes or subject systems about research on the application of new technology in production and daily life.

For the development of agriculture, rural and mountainous areas, the government has created the special application programme in order to transfer new technologies to farmers and rural areas. During this programme, the mobilization of financial support has proved effective in the transfer of new technologies to the local community and farmers.

5.3 INSTITUTIONAL SET-UP OF THE NATIONAL INNOVATION SYSTEM

By the Resolution of the eighth Central Committee of the Communist Party of Vietnam (1996), at least 2 per cent of the annual national budget is to be used for scientific research and technology development. Besides state investments, social ones, mostly from enterprises, have been set up, albeit at a lower level. On an annual average, the non-state investments are about 0.1–0.2 per cent of GDP. The supplemental investments from non-state sectors partly ensure that the overall annual investment in Vietnam was about 0.5 per cent of GDP for the first ten years of the millennium, dropping to 0.4 per cent in 2012 and 2013 (Table 5.2).

In 2002, Vietnam devoted USD 273.5 million (constant 2005 USD in PPP) to R&D. Expenditure for S&T gradually expanded from 0.2 per cent to 0.52 per cent of GDP between 1995 and 2004. In 2012 and 2013, the total state budget for S&T reached USD 627.9 million and USD 673.5

Table 5.4 National budget for S&T activities (percentage of GDP)

Year	2000	2002	2004	2006	2008	2009	2012	2013
National budget for S&T/GDP	0.47	0.52	0.52	0.54	0.52	0.54	0.41	0.39

Source: MOST (2014).

million, respectively (OECD/World Bank, 2014). One specific feature to note is that S&T expenditure is spent not only for R&D, but also construction of facilities and labs, the salary of staff and organization of various S&T-related activities. In 2013, out of 14,144 billion VND (around USD[1] 673.5 million) of state budget for S&T, 6136 billion VND (around USD 292 million) is for facility construction (43.4 per cent), and 8008 billion VND (roughly USD 381 million) is for other expenditure (56.6 per cent). Out of this, 8008 billion VND, only 6088 billion VND from the central government budget was allocated for various purposes like salaries, research missions, programmes and projects.

Thus, the share of actual budget for R&D in total GDP is even lower. R&D expenditure for 2002 was 0.19 per cent of GDP and increased to 0.21 per cent in 2011, which is still very low compared to many other countries (MOST, 2014) but quite moderate compared with other countries in the region. It is the same or higher than the levels in Southeast Asian countries like Thailand (0.21 per cent in 2007), Indonesia (0.15 per cent in 2007) or the Philippines (0.11 per cent in 2007). Of the total, approximately 70 per cent was devoted to applied research and technological development, the rest was spent on basic research. One important feature is that a major portion (about 80 per cent) of R&D expenditure came from the state budget while the private sector and other non-state government institutions did not spend much on R&D activities.

Overall, S&T human resources have increased, but are still lower than what development demands. In 2009, Vietnam had about 3,723,500 persons having university degrees and above who might potentially be engaged in S&T activities (Table 5.5).

More specifically for R&D, in 2011 Vietnam had 134,780 persons engaged in R&D activities, of which 105,230 were researchers, and the rest technicians, supporting staff and others (Table 5.6).

Table 5.5 Personnel with degrees from colleges and above (potential numbers engaged in S&T activities, in thousands)

	1999	2009
Total	1477.8	3723.5
PhD	14.2	24.3
Master	23.3	101.0
Bachelor	1004.7	2650.4
College graduates	435.6	947.8

Source: MOST (2014).

Table 5.6 Researchers working in different organizations (number)

Type of organizations	Total	PhD	Master	Bachelor	College graduates
	105,230	11,501	34,618	55,116	3995
R&D institutes, centres	15,942	1908	4544	9008	482
Universities	52,997	8027	24,337	19,810	823
Semi-professional units	10,165	457	2248	6658	802
Management	9941	369	2170	7091	311
Companies	14,990	508	1115	11,873	1494
Non-governmental organizations (NGOs)	1195	232	204	676	83

Source: MOST (2014).

The quality of the research in Vietnam has been showing certain weaknesses and shortcomings. Because of the unceasing wars and isolation from the international scientific community (mainly in English-speaking countries) for a very long time, the S&T personnel in Vietnam have been confronted with difficulties in updating new S&T trends. There are considerable barriers in exchange of research and international cooperation to be overcome, especially language barriers.

The education and training system in Vietnam is now giving more attention to higher education. But the situation 'excess lectures, lack of workers' has resulted in S&T activities creating only academic and general results which cannot be applied to the realities of production and businesses because they were not in accordance with the requirements of the enterprises.

In addition, the existing pool of scientists and engineers is structured in such a way that most are employed by either the national centres for R&D, universities, ministries or government agencies. Only a small fraction of R&D staff (around 14 per cent) is attached to business firms in Vietnam compared with 35 per cent in China, 81 per cent in the USA, 57 per cent in Japan and Korea (MOST, 2014)

5.3.1 R&D Organizations

Concerning the system of S&T organizations, by the end of 2012 there were 2228 registered S&T organizations (not counting the university system) of which 1074 were public and 1154 non-public. As for R&D organizations, they consist of public organizations, universities and service organizations.

Public organizations

Public R&D organizations belong to academies, ministries and the local government of provinces and cities as well as state-owned enterprises. R&D activities are conducted in most research institutes under line ministries, and there are three national research academies for natural sciences, engineering sciences and social sciences. As mentioned above, there is very modest financing for R&D via the state budget. Vietnam spent around 2 per cent of its government expenditure budget or 0.39 per cent of its GDP on S&T activities in 2013 (MOST, 2014), and its R&D budget was even lower at 0.21 per cent of GDP in 2011 and most publicly funded R&D is conducted in government research institutes. The research infrastructure is below international standards. The research tends to be theoretical, supply-driven and weakly connected to the needs of the productive sector. The national R&D system is still 'organized, financed and managed in such a way that technology transfer is difficult and expensive' (Bezanson et al., 2000, p. 13). Traditionally, public organizations have been privileged in receiving funding from central government to carry out the 'state S&T missions'. These missions are usually organized as research programmes aiming to provide a scientific foundation for policy formulation and the legislative process (social sciences) or to create new S&T outcomes that are significant for economic and social development, for defence and national security, and for human resource development. However, in the absence of an effective mechanism to identify such missions, as well as distribute the research results, it has been argued that the state S&T programmes lack effectiveness. During the last few years, the entire process of identifying, conducting and evaluating state S&T missions has been reviewed in an attempt to identify more efficient and effective procedures. Of all the R&D and engineering organizations in the country, the Vietnam Academy of Science and Technology (VAST) is the largest. By 2012, it had 50 units, of which 34 were research institutes and centres, with nine regional branches operating in various fields of science and engineering. VAST has a staff of more than 4000, including 233 professors and associate professors and 714 wit PhDs. Another state scientific research institution is the Vietnam Academy of Social Sciences (VASS).

In addition to the above institutions, there are several scientific organizations that were set up and managed by line ministries and provinces. These institutions also receive public funding via their ministries/provinces to do research that addresses the scientific/technical and/or policy problems in the fields/areas of the relevant line ministries and provinces.

Non-public scientific organizations

Non-public R&D organizations include: (1) R&D organizations of non-state enterprises; (2) R&D organizations of political, social and professional organizations; (3) organizations belonging to professional associations; and (4) foreign R&D organizations, including joint-venture R&D organizations.

Given the small scale of domestic private enterprises in Vietnam and the weak links with public R&D organizations, the limited non-public R&D activities are mainly conducted in-house by own staff, who may, or may not, have any formal education in scientific or engineering fields. R&D as an organized activity is not common in this sector. With regard to the foreign sector, there are relatively few foreign R&D organizations in Vietnam, and the majority of R&D and engineering work for this sector is conducted in the headquarters of parent firms.

There are numerous so-called R&D organizations associated with political, social and especially professional organizations. However, their interest is mainly in offering consultancy and intermediary services. Few have research facilities able to undertake any significant engineering work.

5.3.2 Patents

In recent years, many efforts have been made to complete the legal framework system for protecting IPR, and to stimulate an increase in inventions and utility solutions (these are the results of research projects that could address problems but are not yet sufficiently sophisticated to be classified as an invention). Applications for inventions and utility solutions has increased threefold in ten years, whereas the number of protection qualifications has remained stable over the years.

Simultaneous with the increased number of applications for protecting IPR, other activities in terms of transfering rights of use and licence start occuring (Table 5.7). In the period 2000–10, contracts for transfering rights

Table 5.7 Contracts for transfering right of use and licence

Contract	2000	2001	2002	2003	2004	2005	2006	2007	2008	2009	2010
Licence contract	80	79	132	99	238	353	135	135	241	163	145
Assignment contract	227	271	266	372	359	168	1200	454	538	648	576

Source: National Office of Intellectual Property of Vietnam (2010).

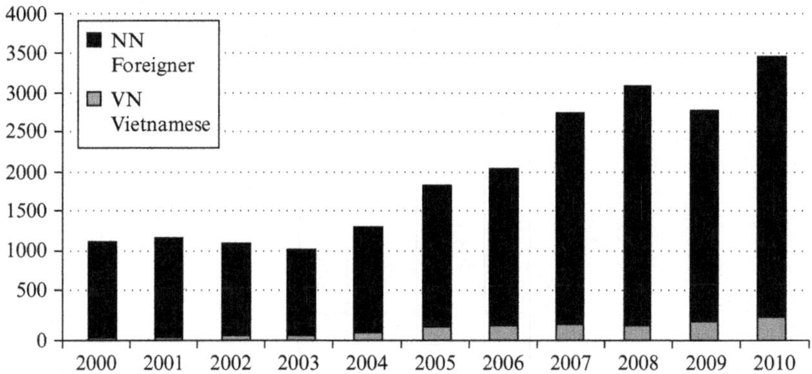

Source: National Office of Intellectual Property of Vietnam (2012).

Figure 5.3 Patent application filed from 2000 to 2010

of use and patent property licence annually increase 20 per cent and 58 per cent, respectively.

Though patent applications and protecting licences have increased, most of them are registed by foreign partners, with a small but steadily increasing part held by Vietnamese as seen in Figure 5.3.

Despite the upward trend in patent applications and registrations in the past decade, Table 5.8 shows the clear continued dominance of foreign registrations. Industrial design patents,[2] on the other hand, are more numerous and mostly obtained by Vietnamese residents, but they usually have a low technological content. The relatively low number of patents may partly be due to lack of capacity to innovate, but also unclear and/or unenforceable intellectual property regimes that dissuade inventors from applying, for fear of a loss of IPR.

5.3.3 Scientific Publications

In 2000, the level of international scientific publications in Vietnam was behind those in Indonesia and the Philippines but since then the level in Vietnam has risen above these countries and was ranked in fourth position in the ASEAN region in 2010 (Table 5.9). International publications for 2011, 2012 and 2013 are 1563, 1929 and 2263, respectively.

Table 5.8 *Registration of invention patents in Vietnam*

Years	1996–2000	2001–05	2006–10	2011	2012	2013	Total
Registration by Vietnamese applicants	161 (2.8%)	482 (6.9%)	1183 (8.1%)	301 (8.2%)	382 (9.7%)	408 (10.5%)	2917 (7.5%)
Registration by foreign applicants	5597 (97.2%)	6543 (93.1%)	13,514 (91.9%)	3387 (91.8%)	3577 (90.3%)	3494 (89.5%)	36,112 (92.5%)
Total	5758	7025	14,697	3688	3959	3902	39,029

Source: MOST (2014).

Table 5.9 International scientific publication of ASEAN countries in 2000–10

Year	Vietnam	Thailand	Singapore	Indonesia	Malaysia	Philippines
2000	357	1420	4247	515	955	484
2001	393	1614	4621	548	1049	405
2002	411	1970	5030	557	1087	531
2003	543	2397	5721	570	1370	559
2004	497	2562	6587	630	1529	565
2005	640	3085	7346	697	1791	706
2006	717	3837	7860	847	2118	716
2007	877	4443	7924	838	2493	766
2008	1068	5246	8904	974	3456	911
2009	1128	6194	9641	1245	4914	1003
2010	1362	6431	10560	1279	6692	1065

Source: ISI Knowledge (2011).

5.3.4 University Research

As mentioned above, although higher education institutions are considered an important part of the S&T system, not all of them are registered as S&T organizations in the statistics. Table 5.10 shows the number of higher education institutions in the academic years from 2010 to 2013.

Almost all universities and colleges fall under the administration of the Ministry of Education and Training (MOET). The Ministry is also responsible for primary and secondary education. In addition, there are two Vietnam national universities (in Hanoi and Ho Chi Minh City). By 2013 (Table 5.10), there were a total of 421 universities and colleges (207 universities and 214 colleges) employing 82,429 lecturers for both public and non-public universities and colleges, of whom 9152 were Doctorates, 36,360 were Masters and 37,243 held university degrees (MOET, 2013).

In general, the government has adopted a policy of creating a selective number of strong public universities as a basis for the further development of universities across the country (the establishment of national universities, regional or community and open universities are just a few examples of such changes to the university system). This move is aimed at creating 'centres of excellence' for university education and research, in an attempt to achieve national recognition.

Table 5.10 Universities and colleges in Vietnam

	2010–11	2011–12	2012–13
Total	414	419	421
Universities	188	204	207
Public	138	150	153
Non-public	50	54	54
Colleges	226	215	214
Public	196	187	185
Non-public	30	28	29

Source: MOET, www.moet.gov.vn.

Traditionally, higher education organizations such as universities and colleges did not undertake research. Instead, they concentrated on the provision of a planned labour force. Recently, although training remains the principal mandate of universities, research has become a more routine activity, with attempts to turn universities into more research-led organizations. Still, only a limited number of university faculties have adequate resources to carry out significant R&D. Human resources for R&D at universities are 62,095 persons, or 46.07 per cent of the total, while researchers working in universities are 52,997 persons (50.36 per cent of total researchers). But finance for R&D in the university sector is only 760.76 billion VND (USD 36.5 million) or 14.37 per cent of the total R&D budget of 5,293.95 billion VND (USD 254.7 million). Public sources account for approximately 15.3 per cent of universities' R&D expenditure. The remaining funding for R&D expenditure comes from contracts with other organizations, of which 29.2 per cent is from enterprises, 6.7 per cent from other organizations and 48.8 per cent from international sources. From 1996 to 2002, universities implemented some 3800 projects and were involved in 90 pilot production projects. Many universities have established their own R&D units. By the end of 2002, within the university sector there were 167 research divisions and 147 centres dealing with technology development and offering consultancy activities. In 2011, MOET, under which many universities work, conducted 6402 projects or 43.88 per cent of the total of all R&D projects by ministries (MOST, 2014).

Despite a large share of R&D projects and numerous efforts, there are several problems with research at universities. Many universities are far from being considered R&D 'centres of excellence'. Universities in Vietnam lack autonomous status. Their staff, especially in public universities, have to face constraints in terms of salary ceilings, human resource management regulations, financial incentives and so on. Despite the move to abolish the separation between teaching and research, there is still a well-documented lack of research and weak linkages between research and teaching (Fatseas, 2010).

The present incentive scheme does not promote a proactive approach among teaching staff in universities. There are few mechanisms to encourage them to interact with other institutions and firms. Cooperation is usually short term, and relies mainly on personal and informal relationships.

In terms of human resources, the number of professors and lecturers is relatively small compared with the increased number of students.[3] Due to overload of teaching, the university staff simply does not have sufficient time left for R&D and/or other learning activities. Ageing staff is another problem, since the majority of professors and associate professors are over 55 years old, with few replacements in the pipeline. During previous years, many scientific and engineering disciplines failed to attract talented young students and as such a shortage of human resources in the university system is foreseen. The low basic salary of academic staff is offered as an explanation for this.

Although the system has seen some recent investment for upgrading, many universities still use equipment and facilities dating from the mid 1960s or mid 1970s. Library systems in many universities are small, and outdated in both their quality and the scope of coverage. The bulk of foreign language literature dates back to the mid 1970s. There has been a lack of electronic links with the national library or investment in centralized information and librarian systems. Moreover, the rate of use is often low due to the poor English capability of the staff and/or their current teaching workload. As a result, teaching curricula are often old, repetitive and lack innovative approaches and new knowledge. During the last decade, the Vietnamese university system has experienced several changes, some more radical than others. First of all, the government paid much more attention to the drive to create a new type of university, which some called research universities, or universities of international level or standard. The government set up a range of partnerships with various foreign partners (France, Germany, UK) to form a number of such universities. These projects received a substantial amount of investment. In addition, a number of private universities come into the scene, with some of them funded by large corporations.

5.3.5 Role of the Business Sector

By 2012, there were more than 600,000 enterprises in Vietnam, of which SMEs accounted for approximately 97.6 per cent (MPI, 2012). As mentioned above, SMEs have contributed significantly to the economic and social development of the country with around 40 per cent of GDP (ADB,

2005; BIA, 2011; Pham Thi Thu Hang, 2011). However, the technology level of SMEs is still outdated. Even large state-owned enterprises do not really pay sufficient attention to the search for updating information, improving S&T knowledge and investing in technology innovation. The rate of technology innovation investment is less than 0.5 per cent of revenue (whereas for India and Korea it is 5 per cent and 10 per cent, respectively).

There are several main causes leading to this situation. Enterprises do not express long-term visions, having little interest in technology innovation for sustainable development.[4] Most enterprises are SMEs and capacities for adopting advanced technological equipment are rather limited.

Only a small part of Vietnam's R&D is conducted in productive enterprises. Recent data from a survey of 325,304 companies show that only 1090 (or 0.34 per cent) reported having R&D activities (MOST, 2014). Innovation in SMEs is concentrated in process development and quality control. Firms do not have sufficient resources for long-term investments, including recruitment of highly skilled labour and equipment to innovate, and technical services are unavailable. SMEs have a hard time obtaining credit from banks. In general, private enterprises are at a disadvantage compared to state-owned enterprises due to lack of access to both preferential and non-preferential government loans.

Findings from several studies (NISTPASS, 1999, 2002) reveal that, in terms of general business services, Vietnamese firms still prefer to perform most services in-house. Firms tend to rely on their internal capacities for engineering, management and marketing.

These studies also revealed certain demand by firms for technology support from the R&D institutes and universities. However, when asked about the sources of innovation ideas, only 10 per cent out of 126 respondents cited R&D institutions and universities as an important source (Nguyen Vo Hung, 2004). There is a demand for technology and training services to be provided by supporting organizations such as R&D institutes and universities. However, the demand has hardly been met; the extent of the relationship of firms with universities and R&D institutions is less than satisfactory.

5.4 CONTRIBUTION OF SCIENCE, TECHNOLOGY AND INNOVATION IN SOME SELECTED SECTORS

5.4.1 Agriculture

According to estimates by the Ministry of Agriculture and Rural Development (MARD), during 1986–2005, the average production value of agriculture in Vietnam increased 5.5 per cent per year. It is thanks to agricultural production that the Vietnam economy can maintain its stability despite the disruption in growth and economic crisis in many economies. In 2013, agriculture contributed nearly 19 per cent of total GDP and one-fifth of total export. Before the renovation period, the share of the population living below the poverty line was 60 per cent, down to 29 per cent in 2003, 19 per cent in 2006, 14 per cent in 2008 and currently less than 10 per cent. In 1985, exports of agricultural products were USD 400 million, up to USD 18 billion in 2010, and reached almost USD 20 billion in 2012 (OECD/World Bank, 2014). Most of the agricultural production has increased productivity and yields due to the application of new crop and production technologies. By applying new technologies in cropping and post harvesting, production of rice, for instance, increased steadily: 40 million tons in 2010, up to 41.8 million tons in 2011, 42.5 million tons in 2012, 44.17 million tons in 2013 and approximately 45 million tons in 2014 (Oryza, 2014). Fishery production also reached a high of 5.1 million tons in 2010 (Nguyen Minh Duc, 2011) and 6.05 million tons in 2013 (VASEP, 2014).

In some large centres of economic development, high technologies have started to be applied in agricultural production. In Ho Chi Minh City, since 2000, the growth rate of agriculture has been high at 3.7 per cent in 2001, 4.7 per cent in 2003 and in 2010 reaching 9.1 per cent. Ho Chi Minh City began to put a high-tech focus into agriculture production by building a 100-hectare agricultural high-tech park. The park uses several new technologies such as hydroponics, deep pond and floating board technology for planting and new cropping; tissue culture for flowers, plants, orchids, fruits and ornamental trees; application of plant regulators; gene technology and enzyme for mushroom production. In the veterinary field, embryonic technology is being used on milk cows and cattle; new technology is being applied to produce bull semen; genetic technologies are being used to produce a new generation of vaccines and stimulate growth in animal breeding; the polymerase chain reaction (PCR) technique is being applied for diagnostic purposes and selection of breeds; information and communications technology (ICT) is being used in managing new species. In the seafood business, thanks to new technologies, fish and other types of seafood have been improved in terms of productivity and quality. In Hanoi, a number of new technology application models have been introduced to enlarge the range of products and increase productivity and yields of crops such as grapefruits, ornamental flowers, organic vegetables. The city has also made efforts to introduce many high-tech-based agricultural parks.

As one of the priorities set up by the Strategy for S&T development, Vietnam gives particular attention to the application of bioscience and biotechnology in the agriculture sector in order to address the social and economic needs for its large population. In this effort, agricultural scientists have implemented artificial insemination, preservation technology, embryo transfer and vaccine production technology. Technologies to form new breeds such as tissue culture and cell generation have been applied widely in the production of rice, hybrid rice, maize and so on. To date, the contribution of biotechnology to Vietnamese agriculture has resulted in self-sufficiency of 25 per cent in the demand for F1 hybrid rice, generating economic benefits of USD 30–50 million per year (MOST, 2003). New rice production has partially contributed to Vietnam's attainment of second position in rice exports in the world. New rice varieties produced by local scientists are used in 70 per cent of rice-cultivated areas. Another key achievement in this field is the provision of over 50 million trees of various types relying on tissue culture technology to meet the demand of forestry development. In genetic preservation and conservation Vietnam has worked on the genetic conservation of animals, vegetation and micro-organism species. These conservation efforts are run by 12 offices and 70 agencies belonging to 8 ministries and branches. Biotechnology has also begun to generate aquaculture species and has been used in preventing disease in prawn species.

Overall, one can say that if Vietnam is to become self-sufficient in food grains and subsequently a net exporter of many agricultural products, the development and adoption of new varieties are of great importance. It is estimated that in the last two decades the introduction of new varieties has accounted for about one-third of the growth of rice production. Public research has played a crucial role. Research in hybrid rice was initiated back in the late 1970s in Vietnam's Institute of Agricultural Science. The development of new rice varieties was later carried out not only in research institutes such as the Cuu Long Rice Research Institute (CLRRI) in the Mekong Delta, the Institute of Agriculture for South Vietnam (IAS) and the Agricultural Genetics Institute in the north, but also in universities such as the Hanoi University of Agriculture, the Can Tho University and An Giang University. However, up to now Vietnamese exports of rice remain, in terms of quality, concentrated at the low end of the international market. Considerable efforts are needed to upgrade the various stages of production and marketing, including through technological innovation.

Vietnam is the biggest exporter of coffee after Brazil and the largest shipper of the robusta beans that are used to make instant drinks. Coffee is second only to rice in terms of the value of Vietnamese agri-product exports. In spite of the importance of coffee as a source of income for large communities and the generation of export revenues, limited resources have been devoted to R&D in Vietnam to improve coffee production and commercialization. The government has sought for some time to improve the quality of coffee exports, including more widespread planting of value-adding Arabica beans and the development of mixed-bean coffees.

The restructuring of the agricultural R&D system under the supervision of MARD has been part of this effort. In September 2005, the number of agricultural R&D agencies under direct MARD control was reduced from 28 to 12. These mergers helped eliminate duplication of research efforts, but do not appear to have solved problems such as the extreme geographic centralization of Vietnamese agricultural R&D (most activities take place in the immediate vicinity of Hanoi and Ho Chi Minh City), which prevents R&D from being closely connected with Vietnam's rural development programmes, and the low productivity of many research institutes, resulting from inadequate incentives combined with lack of an adequate level of human and other resources.

In September 2013, at the National Conference to evaluate the S&T activities under CP Vietnam Resolution No. 7 on agriculture–rural area–farmers, the contribution of S&T to agriculture production was highly praised. It was confirmed that due to R&D activities, 164 new species of plants have been created and put into production, of which 97 are in the foodstuff group, 8 are ornamental flowers, 19 are fruits and 40 are trees of industrial types. Many of the new species have higher productivity and new features including better climate resistance and pest resistance, such as rice, maize, peanuts, potatoes, rubber and particularly new plants used to make furniture.

In one of the studies on innovation system in agriculture (Tran Ngoc Ca and Nguyen Vo Hung, 2012), the status of innovation functions was examined in the context of three specific products of Vietnamese agricultural production: tea; fruit and vegetables; and shrimp. These innovation functions were examined for each product sector along the value chain of production and business, ranging from R&D for seeds and species to planting, harvesting, packaging and sales. The innovation system related to these three products has been mapped and identified two key features: actors/organizations of the innovation system and issues/problems in institutions. Actors and organizations in the innovation system under examination were universities and schools, R&D institutes and centres, value chain organizations like shops, markets, associations from the private sector, various government organizations, law firms and other regulators, enterprise incubators and financial institutions, both public and private. In spite of the achievements of actors, their incompetence was highlighted in terms of several institutional issues and problems. First, there is lack of R&D, testing and training and incubating facilities in many agri-based organizations, reducing their competence. Partly, this is due to limited financial resources, and partly to weak research and training capacity, especially shortage of marketing skills and information. Another emerging issue is lack of will for change and inertia among many actors. In addition, lack of credibility and trust among financial institutions towards innovation is among the most critical barriers for innovation to take place. In general, the policy and institutional environment is not sufficiently conducive and innovative linkages are weak among actors and organizations. As such, several threads points may be raised here. In general, private players are still weak and public organizations and institutions (government) have a more important role to play in promoting science, technology and innovation. Nevertheless, specific features of agri-based production and agriculture in general have different kinds of problems to others in industries and services, such as seasonal specificity of harvesting, productivity of crops and so on. Globalization processes mean that foreign actors are important also in agri-business, be they firms, NGOs or associations.

One interesting observation is that there are differences among various products in terms of innovative behaviours. The same policies are applicable across the agricultural sector, but they seem to have a different impact for various product groups such as tea, fruit–vegetable and shrimp farming. In this differentiation, market factors (both domestic and, more importantly, export) and types of technologies that are being used in the production of these specific products reflect differences – sometimes quite distinct differences.

5.4.2 Healthcare

In the healthcare sector, the Ministry of Health (MOH) is the government organization responsible for all activities concerning taking care of people's health, including preventative healthcare, traditional medicines, drug production and use, cosmetic and food safety and so on. In provinces and cities, the Department of Health (DOH) under the MOH is the responsible agency. By 2010, Vietnam had 1030 hospitals, 44 rehabilitation health centres, 622 regional clinics with 246,300 hospital beds. In addition to the government system, there is a private system of healthcare services comprising 19,895 medical organizations, 14,048 pharmacy organizations, 7015 traditional medicine organizations and five hospitals with foreign investment.

In terms of human resources, Vietnam has a system of medical and pharmacy universities and it produces thousands of medical doctors and pharmacists every year. In addition, numerous schools train nurses, dentists, pharmacists and technicians in the provinces. Currently, there are

250,000 people working in the healthcare sector with 47,000 having higher education degrees. Most medical and pharmacy universities do various kinds of scientific research and technological experiments, using their hospitals as practical experimental organizations. In general, healthcare S&T activities have contributed to some of the latest achievements of the sector in addressing the needs of society. A first notable impact is the ability to control new types of emerging diseases during the last few years. Vietnam, as a tropical country, faces many types of epidemic diseases that require strict monitoring, control and treatment procedures. To detect these new types of disease early and design appropriate measures would guarantee success in dealing with them. The healthcare sector now, thanks to research, is able to detect very new 'strange' diseases the first time they occurr in Vietnam, such as foot and mouth infection in both humans and cattle, avian influenza of H5N1, H1N1 types and so on.

In the treatment of human diseases, many hospitals have been successful in implementing new technologies for complicated operations to replace or transplant organs such as liver, marrow, heart and kidney as well as genetic diagnostic tests. This has saved the lives of many people and increased the capability of the healthcare sector. One example was in 2012, when 37 doctors and nurses in Cho Ray hospital in Ho Chi Minh City (together with colleagues from Korea) successfully transplanted a liver. Similar cases can be observed increasingly in many other hospitals around the country. More recently, stem cell research has been done on diseases affecting blood cells, neurons and bones. This has made it possible for many patients to receive treatment inside the country instead of spending a fortune to travel overseas for these complicated problems. The same can be said for endoscopic operation technology applied to deal with kidney, diaphragmatic hernia, ureter problems and so on.

Among others, biotechnology is one of the priority areas in the application of S&T that has brought a positive outcome to the sector. Over the last few years, artificial insemination technology has been applied widely at some clinics in Ho Chi Minh City and Hanoi. Genetic technology using DNA and protein chips are also being explored in some research institutions and healthcare centres. They have applied fast diagnosis and early diagnosis technology in dealing with heredity problems in newborn babies and the prevention of epidemics such as poultry flu and dengue fever.

The research projects in vaccine production have produced the most significant results in the field of medical biotechnology. Vietnam has mastered the technology and produced nine out of ten kinds of vaccine to offer a comprehensive vaccination programme. This has contributed to the elimination of polio in 2002. Vietnam has recently succeeded in the research and technology development of a new type of class B hepatitis vaccine based on DNA technology (recombinant vaccine). The production of class B hepatitis vaccine since 1997 has resulted in turnover of more than USD 7 million. Four manufacturing enterprises for human vaccines have produced tens of millions of doses per year to support a comprehensive vaccination programme and these vaccines have been exported to India, South Korea and Japan. Two types of vaccines have met with 100 per cent demand and the remainder 40–60 per cent of demand in Vietnam. This helps Vietnam to have almost all the vaccines necessary for its immunization programme. This is especially due to its international collaborations. In addition, genetic engineering, transgenic technologies and GeneCard have been used in the areas of national defence and public security to identify the remains of the dead from national wars, recognize blood relations and identify criminals.

On traditional medicines one can say that, similar to other sub-fields of healthcare, this area also lacked a needed, organized innovation structure conducive to addressing the needs of the population to have affordable treatment and drugs. In spite of many efforts by the government to promote traditional medicines and use of herbal drugs in order to provide more affordable, accessible healthcare, this sub-sector still suffers from disorganized actions and under-resourced programmes (Tran Ngoc Ca, 2012). So far, most S&T activities in the healthcare sector seem to be linked more to modern medicines.

5.5 ISSUES AND PROBLEMS IN CLOSING THE LOOP

There are a number of issues related to how the research community 'closes the loop', that is, aligns its research activities to the national development agenda. To ensure the research activities serve the needs of development, it is essential that, first, the research agenda is set in accordance with development needs. Then, the capability of research organizations to deliver products for development is another issue.

A study was undertaken to examine in depth the capabilities of R&D organizations as well as companies in closing the loop. The cases of ten R&D institutes, universities and firms working in the agriculture and healthcare sectors provide an illustration of the efforts of the research community in support of overall developmental goals. This section provides some experiences of these specific R&D institutes and universities. First, how these organizations set their research mission and agenda is presented. Then, their capabilities are described in pursuing these agendas and achievements so far. Lastly, there is an analysis of the impediments faced by these organizations in aligning R&D activities to development goals.

5.5.1 Research Agenda Setting and Mission

As seen above, the MDGs require that many problems related to agriculture, healthcare, gender, environment and other areas need to be addressed with the help of scientific and technology solutions. The closer the priorities of research projects and programmes to the specific needs of society, the better they serve these needs when the research results become available to end users. In activities of many national S&T research programmes, issues of biotechnology for agriculture and healthcare have been addressed. More specifically, one national programme of S&T development for mountainous and rural areas was designed to solve problems these difficult areas. The end users of research are usually farmers with low income, ethnic minority communities, poor women and children.

Many R&D institutes, university faculties and companies were intentionally created with their mission to provide solutions to development problems. For example, in 1971 the Centre for Maize Research (under MARD) was created and later became the Institute for Maize Research, one of the leading organizations in agriculture research in Vietnam. The institute conducts research on maize and some foodstuffs, ranging from basic surveys to selection of new crops, improvement and application of new production technology solutions for maize and foodstuff production. Most research agendas come from the needs of farmers and at the request of agricultural organizations.

The National Institute of Hygienic and Epidemiology (NIHE), with the international cooperation (from the Swedish Karolinska Institute) was established to develop an appropriate level of research capacity in vaccine research, among other purposes, following instructions from MOH. Different teams in NIHE have obtained expertise in the production of different vaccines. Japanese scientists have trained NIHE staff on Japanese encephalitis vaccine, while Swedish support has focused on a cholera vaccine. From 1991 to 1995, the Ministry of Science and Technology provided support for the institute to carry out state-level research projects on the application of advanced techniques to complete the process of producing an oral cholera vaccine and the technology used to produce Japanese encephalitis vaccine and other vaccines. After publication of the research results in 1996–98, a pilot project was established to expand production of class B hepatitis and oral cholera vaccines with total capital of 2 billion VND (about USD 130,000), giving an investment rate of return of 80 per cent.

Another organization, the National Institute for Medicinal (herbal) Material (NIMM), was created on 13 April 1961 with the function set out by MOH to do research on herbal medicines, test the toxicology of these substances, guide the growth of herbal medicinal plants and train staff working in this area. Its research agenda is a result of a consultation process between researchers, MOH and healthcare workers. The institute has numerous projects working on medicinal plants such as ginseng, and performs research on the localization of imported Chinese species and adapts them for conditions in the northern areas of Vietnam. The institute distributes these new successful breeds to farmers.

As for universities, Hanoi Medical University was established in 1902. The development of HMU led to the formation of two practical hospitals, and now it has three faculties. Training is the most important activity of HMU and it has trained 16,507 medical doctors, 1505 doctor specialists and 143 with public health Bachelor degrees. Since the mid 1990s, its research mission has also been acknowledged as key for the university. To address the needs of society, research activities at HMU are focused on three main research areas: (1) application of new techniques to improve the quality of diagnoses and detect diseases at both onset and molecular level; (2) medical effects of herbs, traditional utilization methods of herbs, molecular bio-production for diagnoses and treatment; and (3) improving public health for workers, poor people, those with genetic abnormalities, people effected by dioxin or toxic chemicals from the war, educational interventions for health protection and so on.

Can Tho University (CTU) was founded in 1966 and includes nine faculties and two research institutes. To support its activities, CTU has developed strong cooperative ties with many international organizations. The main mission of CTU is to train the 16,000 students at CTU and the 15,000 students at associate facilities in other provinces of the Mekong River delta. In addition, there are 1015 postgraduate students in six Doctoral programmes and 25 Master's programmes. The second mission of CTU is research, of which the most notable programmes are applications of new technology in planting, processing agri-products and applied biotechnology research for agriculture. CTU aims primarily to serve the needs of farmers with very low purchasing power, and thus derives limited revenues from commercialization of research.

The Institute of Resources, Environment and Biotechnology (IREB), under Hue University of Agriculture and Forestry, is a subsidiary of Hue University in the central part of Vietnam. It was created as the Centre for Resources, Environment and Biotechnology in 1995. The centre was turned into IREB in 2007, with functions to carry out R&D and technology transfer in respective fields. IREB has two centres: (1) Centre for

Environmental Monitoring and Sustainable Development and (2) Centre for Researching and Transferring Biotechnology. In IREB's research portfolio a range of notable applied research projects can be found such as on three varieties of rice that can withstand drought – having a shorter growth cycle, avoiding the flood season, of better quality compared to previous generations and which are also suitable to the conditions of the region. It is transferring the results to users like agricultural cooperatives and farmers for production on a large scale. The institute also researches the production of asparagus and orchids by tissue culture.

The Central Centre for Research and Production of Herbal Medicines under Ho Chi Minh City Pharmacy University was created in 1987 by the joint efforts of a number of research institutes from Ho Chi Minh City and local government in Phu Yen, a central province of Vietnam. Starting by exploiting the abundant natural plants to produce drugs for export to Eastern Europe, the centre expanded into collaboration with farmers in growing various types of herbal medicinal plants for both domestic and export consumption. Scientists from the Pharmacy University came to work with farmers and provided guidance, teaching and technology transfer to farmers. Products from the centre can also be used for treating diseases of aquaculture species such as shrimp and fish. To conduct these tasks, the centre has expanded its cooperation with a network of other research institutes on aquaculture, chemistry, tropical biology as well as with other universities in Ho Chi Minh City.

The initial origins of some companies were to address development needs. The pharmaceutical company Vimedimex, for example, was formed in 1984 in HCM under MOH. In 2005 it became a shareholding company and is now one of the leading suppliers of pharmaceutical products in Vietnam. In 2002, Vimedimex created a joint venture BV Pharma with the UK Commerce & Technology Transfer Company. The new company specializes in research and production of new products from traditional herbal medicinal plants for both domestic consumption and export. The Future Generation Co. (FGC) Ltd was set up in 1996 for trading and production of tea (black and green) in the domestic and export markets of more than 30 countries. In 2003, the trading company decided to invest in new processed technology for tea production and export and now has two modern facilities in Vietnam with technologies imported from Italy. Recognizing the need for clean production and help needed by farmers in managing their environment, the company works closely with farmers growing tea in specialized tea planting areas on clean production without the use of chemical substances. The aquaculture feed company Tomboy was created in 2001 in Ho Chi Minh City and produced feed for shrimp production. It distributes products for aquaculture production like chemicals, additional ingredients and materials, water fans, other types of equipment and so on. Given the great need to deal with shrimp diseases in the Mekong River, the company has a team of engineers and technicians to help farmers in studying and solving technical problems of shrimp diseases via direct support programmes, such as guidance, publication and distribution of technical papers, brochures, organization of technical workshops and seminars, technology transfer in cooperation with foreign experts, local research institutes and universities. Over the last five years, the company has organized more than 2500 training workshops for all shrimp cultivation zones countrywide for more than 35,000 farmers to learn about technical issues. Tomboy has become a friendly face in helping farmers produce shrimp in Vietnam.

The research priorities, orientation and mission of the above organizations were specifically close to the development agenda. One can say that they were created to deal with the social and economic development needs of society.

5.5.2 Research Capability and Achievements

Having development goals as a mission and research orientation, these organizations also built up capabilities that help them deliver notable results beneficial to the end users including farmers, patients and disadvantaged groups of people.

In 2005, the Maize Reseacrh Institute became a member of the larger VAAS – 80 per cent of all its research products have been accepted by the government as national maize crop products. Since 1993, the institute has supplied around 3500 tons of seeds for other organizations in production and seed trading. It has also patented and transferred seven types of hybrid maize for other Vietnamese and foreign companies. Moreover, it has provided hybryd maize seeds to Laos and Cambodia. Thanks to this research activity, the institute has contributed to the stabilization of sales prices for maize seed crops in the market, and the benefit is estimated at around 100–120 billion VND yearly (around USD 6 million). Every year, the institute receives about USD 600,000 for research and produces around 3000 types of maize breeds. Sale prices of domestically made seeds are 30 per cent cheaper than imported ones and the institute now has 40 per cent of the market share (competing with six foreign companies). As such, the institute has brought great benefits to farmers by replacing imported maize seeds with domestic ones.

On the basis of research results, and recognizing the need in Vietnam for vaccines of various types, NIHE decided to set up a business entity to deal with mass production of vaccines. The research carried out by the researchers of the three laboratories within the framework of international cooperation and later within government research projects was crucial for the company's start-up. The Company for Vaccine and Bio-product No. 1 (Vabiotech) was established in March 2000, comprising a research team from three laboratories. The company's founder is Professor Nguyen Thu Van, who was then the head of the Laboratory of Class B Hepatitis Vaccine. In addition to research results inherited from the institute, start-up facilities transferred from the institute to Vabiotech were also crucial (about 70 per cent of the company's assets). Thanks to the institute's research capabilities and transferred facilities, the results of the research were applied immediately to the production of vaccines. Up to that time, most vaccines available in the market were imported. The sales of vaccine products produced by the company proved that their quality was equal to those of imported products but cheaper in price. Currently, Vabiotech produces and sells six main products, including vaccines for plasma class B hepatitis, recombinant class B hepatitis, inactive class A hepatitis, Japanese encaphalitis, oral cholera and rabies vaccine and other bio-products for diagnosis, treatment and prevention of disease. This helps Vietnam to partly achieve self-sufficiency in vaccine production, essential for community health protection, and saves a significant amount of foreign currency based on import substitution.

Among the notable successes of the CTU in research are the project on rice varieties research (with more than 30 new types of rice having been approved as national varieties), the project on suitable farming systems, technology transfer for pig farming, agri-based food processing, biotechnology for agriculture and the production of various types of aquaculture products like shrimp *Artemia* and new breeds of fish. Supply of services for agricultural needs is also a widespread practice of the university. The university has implemented projects in technology transfer in agriculture and rural development. To commercialize and pursue these activities, CTU has established units like the Company of Plant Protection and Laboratory on Tissue Culture and is working closely with the Department of Science and Technology and local government in several provinces. Farmers, with their small-scale and family-based production, are the main users of innovation and technology transfer from the CTU. These transfers are often done through farmers' groups or following the guidance of local government at the provincial and district levels. The most significant impacts are job creation and formation of new business opportunities. Also, technology transfers from CTU have enabled farmers to combine planting and raising shrimp *Artemia* together with salt production, and to increase crop and productivity in catfish production.

As a result of its research, NIMM has been able to localize more than

200 types of species, of which 37 have been planted on a wider scale. Some of the plants have become very well-known products for treating diseases like malaria (*Artemisia annua L.*) or for the production of vegetable oil (like citronella) for domestic and export markets. The institute also works closely with farmers to help them grow and commercialize products such as mushrooms and various medicinal plants, especially ethnic minorities in the Central Highland or mountainous areas. NIMM has organized research on Crila by tissue culture technology for cancer treatment and a number of other projects aimed at promoting local production and supply of local herbal medicines for poor farmers living in mountainous regions. With the more affordable costs, this should help address the problems of the poor in accessing healthcare services and necessary drugs.

Some activities of HMU address societal needs. HMU has numerous studies on domestic medicine herbs for prevention and therapeutic purposes for local communities. A polyphenol product extracted from green tea leaf is having a restraining effect on lipid metabolism. HMU has also collaborated with a domestic pharmaceutical manufacturer to produce the Uphamorin tablet from the extract of a plant grown in south Vietnam with the effect of strengthening immunity against active chronic hepatitis and nasopharyngeal cancer patients after radio- or chemotherapy. In addition, HMU has been collaborating with a network of hospitals in their training, particularly in the clinical practice for students, and providing online advice to communities via a website on preventing and curing some common diseases.

Among companies, every year BV Pharma (Vimedimex) produces around 100 tons of artichoke, ginger oil extract and other dried substances for treating neuron system infection for export to Russia. The company is also studying how to extract substances from red pine to replace import products in cancer treatment, which costs more than USD 3 million yearly. For this, the company is setting up a Research and Production Centre for growing and harvesting artichoke and red pine, and farmers have been contracted to do this job, under the technical guidance and support of the company.

5.5.3 Problems for Pro-development Research and Related Policies

Despite the fact that many organizations have intentionally set their missions to serve the needs of development, that they followed suitable research agendas and at least some of them have sufficient capabilities to deliver research results, they have not always achieved their objectives. There are clearly some inappropriate policies and government regulations preventing the closing of the loop. This highlights the need for strong government support.

The cases mentioned above also reveal several problems. Most commercialization measures in universities like CTU have to be done via central or local governments, which have budgets for various programmes to support farmers. Nevertheless, it is widely believed that policy has not been supportive enough for the university to play a better role in this agriculture-based region, whether in training or in technology innovation and dissemination. Technology transfers from CTU to the region tend to have immediate but only short-term effects to date.

At IREB, some research projects have been completed and produced results, but it has not found partners to which it can transfer the research result and lacks investment capital for pilot production. There are several issues that hinder the development of R&D at IREB. First is funding. In general, total funding available for R&D at the university is quite limited and MOET provides funding for only five ministerial-level projects at Hue University (the funding at this level is around USD 20,000 each on average). Even worse, at the university level, the funding is only USD 200 per project on average.

The autonomy of R&D institutions is another issue. To date, IREB has relied mostly on its income from external projects. In the context of moving R&D institutions towards greater autonomy, new organizations like IREB may have more problems in facing an uncertain future, without the support of public funds for operation, given biotechnology research cannot generate as much income as activities like environmental services or construction projects.

Within Hue University, there is a lack of structure to deal with the transfer of technology and commercialization of research results, such as a Technology Licensing Office or a Technology Transfer Office. This makes it difficult for research activities to reach the market. University staff lack market and business management knowledge, and thus, even for some useful technologies, it is hard to turn them into marketable products. Universities, such as HMU, also have limited linkage with the private business sector.

In pursuing their goals, companies like New Generation Co. have faced a number of challenges such as lack of application of new technologies in growing tea, which led to the under-supply of clean and good quality raw material for company production. In particular, incentive policies on quality, food safety and hygiene conditions are not in place. There is no single agency responsible for the issue of monitoring the use of chemical fertilizers and insecticides, and lack of related regulations. If there are some, enforcement of the regulations is very weak; the monitoring organizations are in short supply or under capacity. All these considerations point to the need for a better coordinated policy framework enabling good technology companies to better serve the needs of the farmers.

In the case of Vabiotech, although being a firm, it cannot be fully responsible for its staffing and salary system due to its government spin-off organization status. The salary regime is not appropriate, especially given the ceiling salary level imposed by several ministries for all state-owned organizations. The salary regime remains egalitarian, which does not provide incentives for talents. According to the company's director, it is important for her organization to be treated as a real company. This kind of technology-based spin-off organization requires more specific and appropriate policy mechanisms and regulatory framework to be developed by the government.

In Vimedimex, many difficulties are still being faced in manufacturing and processing of medicinal herbs from cultivation, intensive preservation and processing activities. Moreover, this type of cross-sector activity is not under the management of a single ministry, but related to many agencies and therefore the management is quite complex and requires good coordination and cooperation of all relevant ministries that to date are not in place. In addition, there are always issues of policies related to quality and food safety, the management and use of pesticides, chemical fertilizers, monitoring functions, as well as unfair competition like infringement of intellectual property issues and so on. In addition, financial institutions are not very helpful in bridging the needs of industrial production and farmers' needs.

5.5.4 Comments on the Cases

As described above, many organizations, be they R&D institutes, universities or even companies, are working on the issues of serving the needs of farmers, poor and disadvantaged people. Organizations from both the public and private sector (although to a lesser extent) can work on the closing of the loop in development.

Addressing societal needs became one of the missions of these organizations, in addition to other more 'mainstream' functions. Working closely with farmers is one of the most notable modes of operation to make sure needs in agriculture or healthcare are addressed. Still, there are policy and institutional problems that hinder this process. The organizations concerned complain mostly about financial policy and specific funding mechanisms for programmes. To supplement these sources and to avoid the complexity of disbursement procedures, these organizations have to promote their other activities, such as cooperative contracts with users in production sectors; activities to serve the needs of farmers; or increasing work with those organizations that receive funding for the development of key economic zones. In some cases, investment by the state for R&D, technology development and commercialization are not sufficient and very fragmented. Scientists spend a lot of time dealing with procedures and financial paperwork rather than on R&D activities. Many policies are not suitable for the seasonal specificity of agriculture production, which require a long time period before they can produce substantive tangible results. The State Treasury, a monitoring organization responsible for R&D funding, has limited expertise and knowledge on R&D activities and tends to create extra difficulties for scientists. The whole process of agenda setting, selection of projects and monitoring research activities needs to be improved. In general, the gap in knowledge, understanding and misconception between the R&D community and financial management needs to be closed.

Apart from funding from provinces and local governments, international cooperation is another important source of funding that these organizations should further develop. However, procedures and preparation for international cooperation are not always appropriate. The state via various ministries also needs to have appropriate allocation policies in funding R&D activities within the university system. The current situation is considered untenable as the state budget for most university-level research projects is almost negligible, making it practically impossible to do any kind of serious large-scale research that the country needs.

Overall, most R&D organizations and universities do have awareness about the need to address the social agenda and attempt various efforts to pursue these purposes. But in some instances, they are not enough and it depends strongly on the policy framework of the government to bridge the gap between demand and supply of research results.

5.6 SURVEY OF S&T ORGANIZATIONS ON S&T AND R&D ACTIVITIES IN VIETNAM

5.6.1 Introduction

In order to better understand the perception of R&D institutes and universities on the relationship between R&D activities and development needs, a survey was undertaken. The survey team sent out questionnaires in combination with in-depth interviews and case studies to R&D institutes, universities (called Organizations below) and enterprises (called Firms), mainly in the two sectors of agriculture and health. In total, 57 organizations were surveyed, of which 35 are S&T Organizations (including universities) and 22 Firms.

5.6.2 Findings

Most R&D Organizations and Firms, or 91 per cent, were involved in national S&T programmes and projects funded directly by the government. To a lesser extent, they also carried out activities under self-organized projects (69 per cent) or were funded by the foundation mechanisms in a competitive process (37.5 per cent); 59 per cent of respondents were involved in experimental production activities. However, the scope of activities differs between Organizations and Firms. Whereas 80 per cent of Organizations enjoy state-level programmes and 54 per cent receive support from research foundations, the corresponding figures for Firms are 59 per cent and 9 per cent. Conversely, most Firms are involved in in-house R&D and technical change services (95 per cent) whereas a smaller percentage (51.4 per cent) of Organizations engage in this activity.

When assessing the quality of R&D activities within their own area of expertise, most respondents (66.67 per cent) evaluated the R&D quality as average. But Organizations tend to assess R&D quality more positively than firms: 42.86 per cent said it is average, 57.14 per cent considered it good and none characterized it as bad. For Firms, on the other hand, the quality of R&D was considered good only by 9.09 per cent, average by 81.82 per cent and bad by 9 per cent.

Some respondents explained more specifically the choice of 'bad quality of research'. The reasons offered for this were that research results were not up to date regarding the needs of firms; or that research can be good in laboratory conditions but cannot be applied in the reality of the shop floor. Some cited reasons such as outdated research facilities and that results were not credible.

All types of respondents tend to have an average appreciation on the extent to which research results are applied (63 per cent). Much fewer said that they were either very satisfied (14.8 per cent) or not satisfied at all (22.2 per cent) with the level of application.

Again, the satisfaction rate is different among Organizations and Firms. The responses show that R&D Organizations and universities seem to be more satisfied with the rate of application of research results than Firms – 74.29 per cent of Organizations are satisfied, 17.14 per cent are very satisfied and 5.71 per cent not satisfied at all. Meanwhile, 45.46 per cent of Firms are not satisfied at all, 36.36 per cent are satisfied and only 9.09 per cent are very satisfied. This shows the tendency for Organizations that are doing research to value their own research higher than the users – some of

Table 5.11 Assessment on R&D relevance

Overall:				
Criteria	Relevant	Relevant in some projects	Not relevant at all	Others
Total	57	57	57	57
Number	9	40	7	1
%	15.79	70.17	12.28	1.76
For R&D organizations and universities only:				
Criteria	Relevant	Relevant in some projects	Not relevant at all	Others
Total	35	35	35	35
Number	9	24	1	1
%	25.71	68.57	2.86	2.86
For firms only:				
Criteria	Relevant	Relevant in some projects	Not relevant at all	Others
Total	22	22	22	22
Number	–	16	6	–
%	0	72.72	27.28	0

Source: Survey by author.

which are Firms – do. It would seem that research performance is not yet up to the needs of development and business communities.

To clarify how to improve the use of research results, suggestions were offered such as more discussion on research agendas and needs assessment of end users (like firms, communities and so on) before doing the research. Some proposed giving firms/end users the rights to initiate research project and agenda for academic organizations.

In this context, the relevance of R&D activities or the suitability of research results in respective areas to the needs of social and economic development in the country is worth clarifying further. Table 5.11 shows the overall responses on the relevance of the research: 15.79 per cent confirm that it indeed is relevant while the majority (70.2 per cent) find it to be somewhat relevant and a smaller share (12.28 per cent) consider it as not relevant at all. Again, R&D institutes and universities deem their research as relevant or somewhat relevant to a higher degree than Firms do (94.3 per cent versus 72.2 per cent).

To identify irrelevance of research results, a number of reasons were given. First, scientists tend to do what they do best, sometimes not very close to the needs of end users in society. Second, the project implementation sometimes dragged on for too long and even if the initial purpose of a project was relevant, when the projects finished, it was too late and the situation had changed, rendering the results obsolete or irrelevant.

Funding resources for S&T activities of the organizations is one factor that affected R&D implementation. For Organizations, the overwhelming majority (96.4 per cent) get funding from the government budget via state-level projects and programmes; the share of recipients of funding coming from contracts is also high (82.2 per cent). Some R&D organizations and universities receive funding from international cooperation or cooperation with local authorities of provinces, cities or others. Interestingly, only 39.3 per cent of Organizations got funding via foundation mechanisms, which require a higher quality project application in an open competitive environment.

As for Firms only, all got self-funded sources, while only 59.1 per cent got funding from government programmes or projects.

More specifically on channels for funding allocation, the way budgets are allocated affects the relevance and performance of research activities. Although 77 per cent of the respondents hold the view that researchers should propose the research agenda and that it should be up to the Organizations to handle the selection process and competitive bidding for research funding. Such a procedure would provide input for research agenda setting based on what researchers can do, and be decided by organizations but not by researchers themselves. This would make the selection process more objective.

Concerning the relatively new way of allocating state budget via foundation mechanisms (in contrast to direct allocation from the ministerial budget), the majority (57 per cent) considered that state budget allocation should be channelled through foundations. This selection process would follow the whole set of monitoring and management procedures in a fair manner (peer review and so on).

More specifically for enterprises, in order to increase the relevance of research results to the needs of firms, 95 per cent of Firm respondents considered that funds should be allotted to them first, then they would select researchers to conduct research to address their actual needs. It would seem that there is some kind of distrust by firms of the R&D capability of research institutes and universities. At the same time, firms still like to get funds directly from a special programme under the Ministry of Science and Technology (MOST) (86.3 per cent), rather than via other channels

Table 5.12 Preferred forms of funding allocation for firms

Forms of allocation for Firms	Via the National Innovation Foundation	Via channels of relevant ministries or localities	Direct grant from MOST
Total	22	22	22
Number	1	8	19
%	4.5	36.3	86.3

Source: Survey by author.

of ministries or provinces (36.3 per cent). Very few Firms (4.5 per cent) wanted to apply for funding from the National Technology Innovation Foundation (Table 5.12).

In financial terms, there are difficulties faced by Organizations and Firms in conducting S&T projects. First, concerning the adequacy of the funding system for research, almost all respondents, regardless if they come from R&D Organizations and universities or Firms, agreed that it is not adequate – 99 per cent of R&D and university Organizations and 86 per cent of Firms fall under this category. Some respondents elaborated more specifically on this inadequacy. They commonly agreed that funding is not enough for the whole chain of R&D to experimental activities as well as innovation. The problems in overcoming the difficult gap between research results and their commercialization that needs a financial and institutional bridge to cross ('the Valley of Death') tend to be very real. Universities particularly complained that their research funding (either coming from MOET or MOST) was negligible. Some petty project funding from MOET has only symbolic meaning (about USD 200 per project).

On the measures to align research and development goals, nearly half of the respondents (42.11 per cent) said that there were not enough adequate measures to ensure alignment of research objectives and national development goals. However, the reasons for this were quite disparate. In more close-up discussions, some said that a number of good measures has been introduced by MOST in organizing agenda setting meetings, or promoting bottom-up approaches in proposing research needs. At the same time, some said they were not aware of these efforts, while yet others complained that the measures taken were not adequate.

Concerning the links between policy makers and research communities such as opportunities for researchers to present results to policy makers, they were characterized as few, regardless of whether the researchers work at R&D institutes, universities or firms (80.8 per cent did not have any such

Table 5.13 Financial difficulties in conducting R&D

Difficulties	Cannot spend exactly under items of the pre-planned budget	Lack of receipts as requested	Difficult to spend in time as planned	Financial organizations (Treasury) do not understand the specific nature of R&D activities
Total	57	57	57	57
Number	43	35	32	38
%	75.4	61.4	56.14	66.66

Source: Survey by the author.

possibilities at all and only in rare cases (13.5 per cent) did such opportunities exist). Some researchers (5.8 per cent) that have good personal or close contacts with policy makers (sometimes they are policy makers themselves) feel they can always interact with policy makers from time to time. However, no such systematic structure exists for this alignment on a regular basis.

On more specific financial difficulties faced by organizations in conducting scientific research and technology development (Table 5.13), the majority of respondents (75.4 per cent) said that they cannot spend project budget exactly in accordance with items of the pre-planned budget due to the fixed rigidity of the pre-set budget process. Other difficulties were: financial organizations (Treasury) do not understand the specific nature of R&D activities (66.66 per cent); lack of spending receipts as requested by complicated accounting procedures (61.4 per cent); difficulty in spending the project budget in time as planned (56.14 per cent). All these show a gap between the flexible and unpredictable nature of R&D activities and the static and rigid requirement of existing financial procedures.

To address the above financial difficulties, the respondents provided some suggestions. The highest preference for Organizations (56.14 per cent) was to have a lump sum contract for R&D activities, or at least the right to adjust the pre-approved budget flexibly (47.36 per cent), while 54.38 per cent preferred project directors to have the right to readjust the expenditure structure compared to the previously approved total budget line.

Concerning the options to adjust the expenditure structure, the majority among the respondents (35.08 per cent) considered that an appropriate percentage of change compared to the pre-approved budget should be 30 per cent, 21.05 per cent say 20 per cent and only 7.01 per cent would accept less than 10 per cent change.

The survey also provided additional insights related to some specific issues. To understand the significance of research for the individual

Table 5.14 Time spent on doing research

Total	Less than 50%	50–70%	70–90%	Above 90%	Not applicable
57	30	19	3	0	5
%: 100	52.63	33.33	5.26	0	8.78

Source: Survey by the author.

Table 5.15 Freedom to set the research agenda

Total	Always free	Most of the time free	Rarely free	Not free at all
57	3	23	31	0
%: 100	5.27	40.35	54.38	0

Source: Survey by the author.

researcher, one of the indicators was the time that researchers could spend on their research activities (Table 5.14).

Most Organizations or 52.63 per cent (universities are more occupied with a heavy teaching workload) spend less than half of their time on research, while some R&D institutes are more active. Only very few (5.26 per cent) could spend between 70–90 per cent of their time on research. This shows that Organizations have other things to do than research, despite the fact that their mandate is research.

To make sure that research meets the needs of socio-economic development, it is important that the research agenda setting needs to be taken into serious consideration. The freedom in formulating a research agenda is influenced by many factors, and not always guaranteed. According to responses provided (Table 5.15), in most of the cases researchers are more or less free (40.35 per cent) or have some limited freedom (54.38 per cent) in selecting what kind of research they do. There have always been some factors influencing their research agenda. Only in a few cases (5.27 per cent) is there total freedom and none of the researchers felt there was no freedom at all.

Respondents also provided answers to what factors influenced them the most while formulating research projects (Table 5.16). Because respondents can have several priorities at the same time to consider, the table shows only the tendencies of impact by these priority factors. Most influential factors are priorities of funding agencies (31.6 per cent) and faculty research priorities (26.3 per cent). National development goals are

Table 5.16 Factors influencing research agenda setting

Factors	National development goals	Faculty/department research priorities	Priorities of funding agencies	Academic impact
Number	12	15	18	12
%	21.1	26.3	31.6	21.1

Source: Survey by the author.

sometimes more general and less specific, and therefore it is more difficult for researchers to count them as influential or not (21.1 per cent). These data reveal the fact that research project agendas are not always influenced and linked to socio-economic development issues, but shaped more by the very specificity of related organizations, departments, faculties and general academic consideration, which, in turn, depends on the specific orientation of individual researchers.

For research to better serve social and economic development, dissemination of research results is important. The majority of respondents (55 out of 57 or 97 per cent) believed that their research produced some results and had an impact either on the local or national level. This is in resonance with patterns where respondents considered their research to some extent relevant to the needs of society. However, one should note that it is difficult for researchers themselves (either in R&D organizations or firms) to confirm that their projects or products are not relevant (having no impact at all). To them, one way or another, they should have some impact, if not at the national than at the local level.

Concerning the specific impact, a majority of respondents (38 out of 55 responding to the question, or 69.1 per cent) categorized their results as having social impact. It is not easy to calculate economic impact immediately on completion of the research. In many cases, it is not possible at all to estimate economic gains of the research projects. This was true even for projects that produced very tangible results for industry and agriculture production – 31 per cent responded that they felt a project had resulted in an economic impact and the remaining 9 per cent saw an impact on policy.

The links between research projects and policy makers are underdeveloped. Only part of the surveyed projects (20 per cent) were known to policy makers at the time of execution. Moreover, these projects may be known to officials at lower levels of local government agencies, but not at the central government level. It was quite difficult for projects in the natural and engineering sciences to be known to policy makers who tend to be more familiar with social science projects.

One solution to bring research closer to policy making is to engage more policy makers. But only in a handful of projects (2 out of 57 responses) was there a policy maker in the research team. These persons also come from local policy-making circles. And again, these tend to be social science research projects related to some policy-making process itself (economics, social studies and so on at the local level).

In answering the question of what the problems are in transmitting research results to policy makers, most responses highlight lack of channels for communication (87.72 per cent) and lack of tradition in cooperation (66.67 per cent) (respondents called this a lack of tradition or habit of translating results to policy makers). When the research is finished, the main recipients of the results are the ones who manage, monitor or fund the project and to whom researchers feel obligated to report. There was no tradition to transfer these results beyond these managers to policy making.

Responses to the question on how to improve the interaction between the research community and the policy-making community include:

- increase communication opportunities between research groups and potential end users in policy-making circles
- create specific channels for translating research results like forums, workshops, roundtables and so on
- train researchers on how to turn their research results into policy products like how to write a policy brief, policy reports, organize policy dialogues
- set up some commonly used policy terminology or mutual understanding between research groups and policy makers
- formulate requirements for researchers to turn research results into policy advice.

5.7 CONCLUSIONS

As demonstrated from the survey results and case material, there are some discernable patterns in the efforts to make research closer to the needs of development. Many R&D organizations and universities in Vietnam were created to serve the needs of development and some were even specifically set up for the purposes of agriculture, healthcare or other needs. As such, their research mandate or training specialization is close to the development agenda. The cases of selected organizations also reveals that they were able to deliver what they were expected to do in terms of producing research results and training on the needs of development in agriculture and healthcare. In some instances, the achievements were quite remarkable like the self-sufficiency of vaccines for the immunization programme and new species and crops in rice and maize production.

However, there are a number of problems to be noted. The quality of research in many research institutes and universities is still poor. There is a gap between what can be done under laboratory condition and the application of new research results on a massive scale of production. In some cases, research results were deemed not credible. Firms as end users of research results tend to look down on this performance.

How to create a relevant R&D agenda is the next problem. Research communities are not always free to set up an agenda and many view R&D relevance as not very high. Some reasons were given in the survey as to lack of relevance of research results. Scientists tend to do what they can do best, sometime not very close to the needs of end users in society. Project implementation sometimes has a long time lag to when the projects finishes, and the results are no longer relevant in a fast-changing environment. Due to the fact that development goals are sometimes too general, R&D agendas are set up not entirely to meet development goals but by the specificity of related organizations, especially those who fund the research. This partly prohibits the freedom of researchers in agenda setting research.

There are several issues concerning the policy framework that contribute to this situation. First, financing for R&D is seen as inadequate in many cases as well as from the perspective of organizations participating in the survey. One of the problems that emerged is lack of suitable funding to overcome the gap (Valley of Death) from R&D to market commercialization. Further, financial management mechanisms such as paperwork and other complicated accounting procedures hinder the process of conducting R&D. The Treasury, as the organization dealing with financial management of R&D activities, is named as one of the bottlenecks. Overall, the view from the cases and the survey also points to a need for specificity of policy and regulations. To address this problem, many respondents suggested new allocation channels via research foundations and emphasized the role of the foundation mechanism.

Among issues for closing the loop between R&D and development goals, alignment of the research agenda with development goals is crucial. So far, the view from relevant organizations is that there are not enough measures to close this gap, despite some recent new initiatives such as dialogue, conference and forum. There are still few policy links between the research community and policy makers. There are little contacts between them, and even if occasionally some are organized, like policy dialogues, there is still no systematic structure on regular basis in place. Moreover, there is little direct engagement of policy makers in the research process.

The last problem is the lack of communication channels and cooperation tradition between researchers and policy-making circles. There is still hesitance about sharing research results among research communities, and between them and policy-making organizations. The culture of involvement by policy making in R&D processes does not seems a prevalent mode of interaction.

This chapter has provided background to the science, technology and innovation system in Vietnam. It has highlighted that the system has contributed to many achievements of socio-economic development of the country in the last few decades. Many activities have been tried to make S&T better serve the needs of development. At the outset, the policy and strategy of Vietnam was to create specific research organizations for development, especially in agriculture and healthcare, which are among the MDGs' priorities. These organizations have indeed done quite a good job, with some important deliveries of research results, leading to concrete results in agricultural production, healthcare and social causes overall. This, to a certain extent, contributed to the achievement of many MDGs in Vietnam by 2010, such as poverty reduction. S&T activities have been undertaken to deal with the needs of the poor and disadvantaged people, in particular, those in mountainous and rural areas. There have also been attempts to improve the linkage between R&D and development by organizing a number of policy consultancies, or dialogues to improve the setting of research agendas. But many problems still exist, most of them policy related such as lack of adequate financial resources and management mechanisms, or lack of communication and dissemination channels for research and policy, to name a few.

In this context, there is a need to adjust the policy framework to make R&D closer to development. A numbers of policy measures could be explored to improve this situation such as setting up specific agendas for R&D that answer the concrete needs of development, rather than general themes. Increased communication and dissemination channels between research groups and potential end users in policy-making circles could help to make research closer to reality. Better facilitation of the commercialization process could also address the difficulties in conducting the third mission of research organizations and universities while turning research results into end products or services for the benefit of society.

NOTES

1. Exchange rate for 2013 is around 1 USD = 20,781.6 VND.
2. An industrial design is a product's shape that is formed by lines, three-dimensional form and colours or a combination thereof, and which is a worldwide novelty and used as the

pattern of industrial or handicraft products (Article 784 the Civil Code). The protection title is valid for five years.
3. One professor has to teach about 30 students, while in other countries this ratio is about one professor for 15 students.
4. According to the Central Institute for Economy Management (CIEM), enterprises are only interested in tax policy (36 per cent), credit support policy (39 per cent) and pay much less attention to technology transfer policy (12 per cent).

REFERENCES

ADB (2005), *Vietnam: Private Sector Assessment. Strategy and Program Assessment*, Manila, available at http://www.adb.org/documents/viet-nam-private-sector-assessment (accessed 23 June 2016).

Bezanson, K., Tran Ngoc Ca and G. Oldham (2000), *Science, Technology and Industry Strategy for Vietnam to 2010*, Hanoi: UNIDO/DSI.

BIA (2011), 'SME in Vietnam', *Business in Asia*, available at http://www.business-in-asia.com/vietnam/sme_in_vietnam.html (accessed 25 October 2015).

CIEM (2010), *Vietnam Competitiveness Report*, Hanoi.

Fatseas, M. (2010), 'Research–industry cooperation supporting development in Vietnam: the challenge of translating policy into practice', in G. Harman, M. Hayden and Thanh Nghi Pham (eds), *Reforming Higher Education in Vietnam: Challenges and Priorities*, Vol. 29, Springer Higher Education Dynamics, Netherlands: Springer pp. 103–15.

GSO (General Statistics Office) (various years), *Annual Reports*, Hanoi.

MOET (2013), *Annual Statistical Report*, Ministry of Education and Training, available at http://moet.gov.vn (accessed 25 October 2015).

MOST (2003), *Review of the Biotechnology Development in Vietnam*, Ministry of Science and Technology, Report, National Conference on Biotechnology, Hanoi.

MOST (2014), *Vietnam Science and Technology 2013*, Ministry of Science and Technology, Hanoi.

MPI (2012), *Report on Socio-economic Status for the First 7 Months of 2012 and Implementation of Resolution 01/ND-CP*, submitted to regular government meeting, Ministry of Planning and Investment, July.

Nguyen Minh Duc (2011), *Value Chain Analysis of Fishery Production in Vietnam*, Ho Chi Minh City: Nong Lam University.

Nguyen Vo Hung (2004), 'Innovation survey of domestic firms', Project Working Paper, NISTPASS.

NISTPASS (1999), *Technological Capability of Firms in Economic Sectors*, Final report of the survey of six economic sectors, Hanoi.

NISTPASS (2002), *Survey of the Supply Capability of Organizations in the Technology Infrastructure*, Final report of the survey of R&D and technical service organizations, Hanoi.

OECD/World Bank (2014), *Science, Technology and Innovation in Vietnam*, Paris: OECD Reviews of Innovation Policy.

Oryza (2014), *Vietnam Rice Production*, 24 September, available at http://www.oryza.com/tags/vietnam-rice-production (accessed 25 October 2015).

Pham Thi Thu Hang (2011), *Vietnam Enterprises Annual Report*, Vietnam Chamber of Commerce and Industry (VCCI), Hanoi.

Tran Ngoc Ca (2012), *Mapping Traditional Medicines Innovation System in Vietnam*, Unpublished report, Hanoi: NCSTP.
Tran Ngoc Ca and Nguyen Vo Hung (2012), *Toward an Evolving Innovation System in Agriculture. Case of Three Products: Vegetable/Fruits, Tea, and Shrimp in Vietnam*, Hanoi: Science and Technics Publishing House.
VASEP (2014), *Vietnam Fish Production in 2013*, Vietnam Association of Seafood Exporters and Producers, available at http://www.seafood.vasep.com.vn/Daily-News (accessed 25 October 2015).
World Bank (2013), *Statistical Country Profiles*: *Viet Nam*, Washington, DC.
World Bank (2015), *Vietnam Data*, available at http://www.data.worldbank.org/country/Vietnam (accessed 25 October 2015).

6. The national innovation system in Tanzania and its relevance for development

Bitrina Diyamett, Heric Thomas, Lanta Daniel, Justine Liberio and Carlos Aguirre-Bastos

6.1 INTRODUCTION

Ideally, research outputs lead to the build-up of a knowledge pool from where they are drawn to inform policy and/or be applied to social, economic or environmental developments. In this book, 'Closing the Loop' is an approach that has been used to analyse how far the outputs produced or adapted by the national innovation systems are actually used by policy makers and/or socio-economic agents in developing countries. In most developing countries the 'Closing the loop' process is poorly understood or applied, although it is rhetorically claimed in all national science, technology and innovation policies.

This chapter analyses the case of Tanzania. The study focuses on the agriculture and health sector systems of innovation, as they contribute particularly to inclusive development.

Section 6.2 presents a brief overview of the socio-economic context of the country, emphasizing the issue of poverty reduction. Section 6.3 provides a more detailed description of the national innovation system and its present status, including the policies and strategies that have been adopted for their promotion and strengthening as well as how the research agendas are set. Sections 6.4 and 6.5 describe and analyse the performance of the sector innovation systems in agriculture and health, respectively. For the analysis and in terms of methodology, both primary and secondary data were used. Primary data were collected through two surveys that included the agricultural and health universities and research and development (R&D) institutions of the two sectors, whereas secondary data were obtained through reviewing relevant literatures. The results of the survey are discussed in Section 6.6. The chapter concludes with a discussion on the implications of the survey results for policy decisions.

6.2 THE SOCIAL AND ECONOMIC CONTEXT

6.2.1 General Overview

As a consequence of economic reforms, solid macroeconomics policies and favourable external conditions, the Tanzanian economy has experienced in the last 15 years or so an important and robust economic growth and strong resilience to external shocks, driven by fast-growing and relatively capital-intensive sectors. Gross domestic product (GDP) growth averaged 6.3 per cent from 2008 to 2013 (World Bank, 2015a). Continued donor assistance supported this positive growth. The country is among the largest recipient of foreign aid in the world. External resource flows have increased from about USD 1.1 billion in 2000 to about USD 2.6 billion in 2013. Of the latter figure, USD 2.2 billion was channelled through the government while USD 440 million was disbursed to non-governmental organizations (URT, 2013a), mainly provided in the form of direct budget support.

The volume of foreign direct investment (FDI) has also been increasing since 2005, averaging USD 603.3 million annually. The *World Investment Report* 2013 (UNCTAD, 2013) highlights a significant increase in FDI in 2012 up to 38.77 per cent from 2011 (USD 1229.4 million to USD 1706.0 million). This latter growth was driven by primary sectors, such as oil and gas exploration. The largest share of FDI inflows went to mining and tourism. In spite of this increase in FDI, the country has seen many of its advantages rather limited due to the low level of human resource development in terms of quality and skills.

Moreover, the contribution of the large inflow of FDI to local technological capability is minimal largely because investments are geared towards sectors that have poor linkages with the rest of the economy. Even for the sectors with potential linkage such as manufacturing, the linkage is weak; there is appreciable technology distance between the local firms and FDI (Diyamett et al., 2011), providing minimal opportunity for learning and upgrading.

A main contributor to the GDP is the agricultural sector that provides over 60 per cent of exports and employs 80 per cent of the workforce. In the period 2007–13 it represented 27–32 per cent of the GDP. The above contribution notwithstanding, the sector has exhibited relatively sluggish growth that averaged about 4 per cent between 2005 and 2008, although it picked up to 5.3 per cent in 2013 (URT, 2013a). As recognized by the *National Strategy for Growth and Reduction of Poverty*, which is discussed further in this chapter, reasons for such growth behaviour include: 'poor infrastructure to support agriculture, inadequate extension services, poor

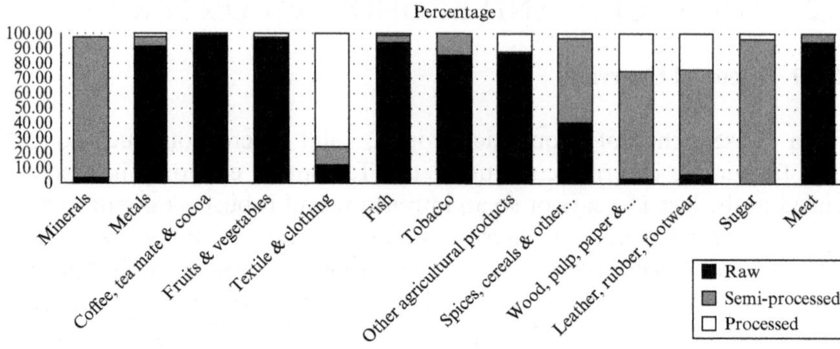

Source: UNIDO (2012).

Figure 6.1 Composition of resource-based product export in Tanzania

production technology, low value addition, lack of appropriate financing mechanism for agriculture, absence of reliable market and prices and environmental degradation' (URT, 2005).

The country is generally rich and diverse in natural resources but faces important environmental challenges. One sector of growth potential is in the fisheries sector (URT, 2013a). Environmental challenges include soil degradation, deforestation and destruction of coral reefs. Wildlife is threatened by illegal hunting and trade. It is also estimated that climate change will be significant. The World Bank notes, however, that government has been taking steps to overcome this situation (World Bank, 2009).

In the industrial manufacturing sector and, according to the United Nations Industrial Development Organization (UNIDO) *Tanzania Industrial Competiveness Report* (UNIDO, 2012), as shown in Figure 6.1, Tanzania continues to rely heavily on unproductive agricultural sectors, the extractive sector and low value added manufacturing. It is emphasized that the performance of the manufacturing sector is not impressive as it lags behind in both quality and quantity of goods produced and exported. The sector remains relatively small, with most activities concentrated on the creation of simple consumer products such as foods, beverages, tobacco, textiles, furniture and wood allied products. In spite of its size, the manufacturing sector continues to be of considerable importance to the Tanzania economy as it is still one of the most reliable sources of government revenue in terms of corporate and income taxes, accounting for over half of the annual government revenue collection. The contribution of the manufacture sector to the overall GDP of the country has averaged 10 per cent (URT, 2013a). Over the last decade, activities within the sector have been registering an annual growth of over 4 per cent and the sector is currently the third most important to the Tanzania economy after agriculture and tourism (URT, 2011). Despite this relatively good performance, the sector is affected by constraints such as the high cost of doing business.

One other important growth sector is tourism, as the country has some of the world's finest tourist attractions and game reserves and there are large opportunities for further expansion. Today, tourism represents over 50 per cent of GDP.

The telecommunications sector has also seen an important growth in the past years, following the adoption of new policy, which is considered among the most liberal in Africa, and technological advances such as the installation of the fibre optic international submarine cable, and the introduction of the 3rd and 4th (3G and 4G) mobile services and wireless broadband networks.[1] As a result, by 2015 BuddeCom considers that the penetration rate for mobile communications and the internet will have reached 75 per cent and 26 per cent, respectively, a significant level when compared to other African countries.

A general contribution to the GDP in the service sector comes from those subsectors with less employment potential and low skills such as retail trade and repair, largely belonging to the informal sector (Mutambala et al., 2013). The turning point for the high contribution of the low skill and low employment potential to the GDP of these service subsectors was a result of the privatization exercise of the mid 1990s that displaced labour from formal employment.

Another important feature in the economic structure is the negative trade balance. In 2011 the level of exports in Tanzania was USD 5.098 billion while the level of imports was USD 9.828 billion (Mundi[2]). In 2012 the level of exports in Tanzania amounted to USD 5.997 billion while the level of imports amounted to USD 10.33 billion. Moreover, exports are largely based on low value added natural resources (Figure 6.1).

In social terms, in spite of the observed economic growth, Tanzania is in the bottom 10 per cent of the world's economies in terms of per capita income. The GDP per capita (at purchasing power parity, PPP) is estimated at USD 2270 in 2012 and 2430 in 2013.[3] It is clear that poverty conditions have improved, suggesting a stronger impact of economic growth on poverty reduction, but the response to this growth has been very slow and it is particularly pervasive in the rural areas where around 70 per cent of the population lives. In these areas about 10 million people live in poverty, and 3.4 million live in extreme poverty, compared to less than 1.9 million living in poverty and 750,000 people in extreme poverty in the urban sector (World Bank, 2015b).[4]

One trend that is becoming important is the so-called endogenous development policies, emphasizing the use of natural resources, local infrastructure and human capacities that are available at the national level (Aguirre-Bastos and Rebois, 2011).

6.2.2 The Development Plans

Between 1995 and 1999, the government adopted a *National Development Vision 2025* (URT, 2000), understood as 'an articulation of a desirable future condition or situation which a nation envisages to attain and the plausible course of action to be taken for its achievement'. The Vision is said to bring out the feelings and desires of the Tanzanian society. By 2025, 'the economy will have been transformed from a low productivity agricultural economy to a semi-industrialized one led by modernized and highly productive agricultural activities which are effectively integrated and buttressed by supportive industrial and service activities in the rural and urban areas.'

The Vision sets the goal for the country to have transited from a least developed country to a middle income country by the year 2025 with a high level of human development when a solid foundation for a competitive and dynamic economy with high productivity will have been laid (see also URT, 2010, 2012a).

Of relevance to the discussion in the chapter, the Vision considers capacity building in science, technology and innovation (STI) as the driving force that enables basic needs to be met by increasing productivity and competitiveness. This emphasis on STI has been placed in a number of national development strategies and programmes.

Under Vision 2025 and given the results of the earlier strategy for poverty reduction, government established MKUKUTA II: *National Strategy for Growth and Reduction of Poverty* (2010/11–2014/15) (URT, 2010) as an organizing framework set to define the efforts that are necessary to accelerate the reduction of poverty. The Strategy is built on four key fundamentals, which have elements that pertain directly to the performance of the innovation system, such as the efficient use and development of factors of production, including human capital/resources, the strengthening and establishment of well-functioning institutions and markets and ensuring good economic governance.

MKUKUTA has given high priority to eradicating extreme poverty and promoting broad-based growth. The World Bank report, *Tanzania Mainland Poverty Assessment* (World Bank, 2015b) considers that MKUKUTA has provided the correct basis for the observed improvement in poverty reduction. The adopted Strategy recognizes that human resources remains one of the critical constraints in its effective implementation and the execution of sector policies, strategies, programmes and projects. It also recognizes the need to transit to a knowledge-driven economy, and highlights research as an important instrument that urgently requires special attention together with drivers of growth such as advanced human capital.

MKUKUTA calls for investment in R&D to be driven by a need to address a given practical problem of a technological, commercial, economic, social, as well as environmental nature. This purpose can be equated to a call for the accelerated development of social innovations. To develop a strong national system of innovation that can actually produce innovations, the links and integration between research, policy and productivity, particularly in the manufacturing, agriculture and trade logistics sectors are recognized as critical.

Tanzania is near to concluding the implementation of the Five Year Development Plan (2011/12–2015/16) (FYDP), intended to implement Vision 2025 in view of the new paradigm it establishes. The present Plan has aimed to unleash the country's resource potential in order to fast track the provision of the basic conditions for broad-based and pro-poor growth; and considers development of the national system of innovation to be of strategic importance in this regard (URT, 2012a). An assessment of this purpose after the Plan is completed will show that its provisions have been partially fulfilled.

6.3 THE NATIONAL INNOVATION SYSTEM

6.3.1 Actors in the National Innovation System of Tanzania

The innovation system approach has been adopted in Tanzania for quite some time (for a detailed discussion, see Chapter 3, Section 3.4). Figure 6.2 provides a simplified view of how the Tanzania national innovation system is organized. The leading institutions of the system are described below (Aguirre-Bastos and Rebois, 2011).

The Ministry of Communication, Science and Technology (MCST)
MCST is the head of the science and technology (S&T) system as a policy-making body. Other ministries are also in charge of overseeing R&D activities in their respective sector. The Ministry was created after an institutional reform that abolished the Ministry of Science, Technology and Higher Education.

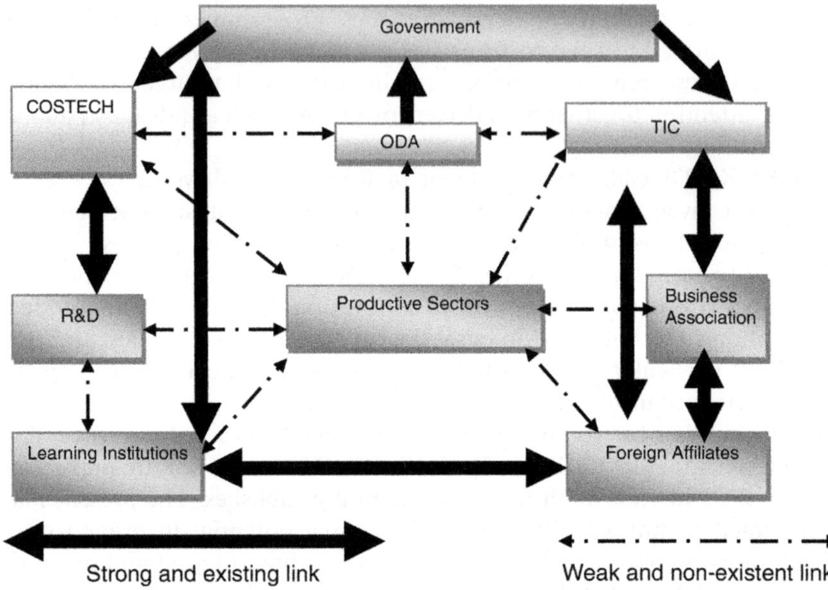

Source: UNCTAD (2002).

Figure 6.2 The structure of the Tanzanian national system of innovation

The Commission for Science and Technology (COSTECH)

COSTECH is a parastatal organization, established in 1986, now under the MCST. Its main objectives are to coordinate and promote S&T development activities. It is also the main adviser to the government. The Commission has established standing committees (R&D Advisory Committees) that act as the organization's think tanks on different issues. At present, there are committees in agriculture and livestock; natural resources; industry and energy; public health and medical research; environmental research; basic sciences; social sciences; development and transfer of technology; and biotechnology.

In addition, COSTECH serves three other national technical committees: Tanzania Award for Scientific and Technological Achievements (TASTA), National Award for Research in Science and Technology (NARST) and National Award for Environmental Management (NAEM).

Under COSTECH operate the National Research Registration Committee and the National Fund for the Advancement of Science and Technology. Internally, COSTECH operates through the Directorate of Research, Coordination and Promotion, the Directorate of Information and Documentation, the Directorate of Administration and Finance and the Centre for Development and Transfer of Technology.

The National Fund for the Advancement of Science and Technology was established under COSTECH in 1995, as an inter-ministerial fund channelled by the Treasury through the MCST to provide resources for a large set of scientific and technological activities, R&D, technology development, prize awards, support meetings, school science fairs, scientific expeditions, publications, writing of textbooks, assist in the acquisition of scientific equipment, support scientific and engineering societies, provide fellowships for training and enable participation in meetings. The Fund has historically received a limited amount of resources, except in 2009 when it received 30 billion TAS[5], which served to finance some well-qualified research projects. The government committed to apply 1 per cent of GDP to R&D as of 2009/10, but this goal was never reached.

The Tanzanian Academy of Sciences (TAAS)

The Academy (TAAS, 2007) is a learned non-governmental organization established in February 2004 and registered on 13 April 2005 with the main objectives of 'promoting S&T learning and the utilization of S&T knowledge for national socio-economic development, as well as to provide a forum for the discussion of various issues pertaining to science and technology'. The Academy is made up of 25 founding members, and has conducted a limited number of activities. Because of its membership it has the potential to contribute to the operation of the national innovation system if the appropriate internal and external means can be found.

Other actors

The national innovation system comprises other key actors: Centre for Agricultural and Rural Technology, Tanzania Bureau of Standards (TBS), Tanzania Engineering Manufacture and Design Organization, Tanzania Industrial Research and Development Organization (TIRDO) and Tanzania Industrial Studies and Consulting Organization (TISCO). It also includes universities and technology institutes and business associations: Tanzanian National Business Council (TNBC), Tanzania Chamber of Commerce, Industry and Agriculture (TCCIA), Confederation of Tanzanian Industries (CTI) and Tanzanian Private Sector Foundation (TPSF).

Other important actors are the Tanzania Investment Centre (TIC), Overseas Development Agency (ODA), foreign affiliates such as donors or development partners. According to the United Nations Conference on Trade and Development (UNCTAD, 2002), and many other studies, there are very weak linkages among the actors in the Tanzanian national system of innovation.

6.3.2 The Universities as Key Actors of the National Innovation System

Education in Tanzania

During the past years, Tanzania has made important inroads into the improvement of primary and secondary education, through the adoption of the Education Sector Development Programme (ESDP) (URT, 2008)[6] initiated in 1997 and revised and updated in 2001 and 2008 (covering the period 2008–17).

The *Medium Term Strategic Plan: 2006/07–2009/10* (URT, 2006) of the Ministry of Education and Vocational Training (MOEVT) is designed to articulate the government's vision for the development of Tanzania's education and training system. It is a conscious process in which MOEVT assesses its current state of affairs and the likely future condition of its environment under which it will operate. Thus, the *Medium Term Strategic Plan* (MTSP) provides strategies and sets specific priorities for the sector. The key areas that have been identified are: enrolment, access and equity, quality, efficiency in education management, cross-cutting issues and education management information system. Accompanying the new policies and strategies, spending in education has been maintained at a high level in the past decade, the percentage of government spending on education as part of its total spending has fluctuated between 15 per cent and 20 per cent in the period 2005–15.[7]

Universities make up an important element of the national innovation system. All the system's stakeholders in Tanzania agree that it is necessary to increase the number and the quality of human resources in both the public and private sectors, that it is necessary to increase the public budget destined to the universities, technical colleges and vocational schools and that it is necessary to increase their infrastructure capacities. These three large and fundamental needs represent the base on which it is possible to improve and enhance the performance of universities in the innovation system (Aguirre-Bastos and Rebois, 2011). It is a political decision that will support all other actions that can take place.

Considering the need to have highly skilled and well-trained human resources, in March 2010, MOEVT put forward its proposal for the adoption of a *Higher Education Development Programme* (HEDP) (MOEVT, 2010). The Programme identified some key interventions that had arisen from the issues recognized in two previous studies sponsored in 2002 by the then Ministry of Science, Technology and Higher Education on post-primary education and training and post-secondary education and training and the respective sub master plans adopted in 2003, one on higher and technical education (2003–18) and the second on S&T (2003–18). The development of HEDP is geared towards making these two sub master plans operational. HEDP also takes into consideration regional and international conventions and protocols on education of which Tanzania is a signatory.

The primary goal of HEDP is to make a positive contribution towards the achievement of goals stipulated in Vision 2025, MKUKUTA and MKUZA (for Zanzibar), and the Millennium Development Goals (MDGs). By 2012 interventions started to be applied on a limited basis but are yet to be completed by 2016. The main goal of the HEDP is expected to be achieved through three developmental objectives:

a. Establish a comprehensive and coordinated higher education system by undertaking institutional reforms.
b. Improve delivery of higher education through enhancement of relevance and diversification of curriculum, increased access, equity and quality.
c. Enhance the capacity of the higher education system in order to maintain and sustain all its functions effectively and efficiently.

This programme is meant to cover universities and university colleges accredited by the Tanzania Commission for Universities (TCU) and higher education service agencies. For non-university higher education institutions accredited by the National Council for Technical Education (NACTE), there is another programme being developed, the Technical and Vocational Training Development Programme. For the purpose of HEDP, higher education refers to the education offered by institutions registered by TCU and does not include education offered by institutions that are registered under NACTE.

Higher education in Tanzania has suffered enormous changes in the past decade. In 1995 there were only three public universities, while today there are 32; eight public universities, three public university colleges, 12 private universities and nine private university colleges, all with a total student population of around 117,000.

The situation of Tanzania's higher education system has been analysed extensively by many authors (see, for example, Aguirre-Bastos and Rebois, 2011 for a bibliography). Specific weaknesses and challenges have been identified by HEDP; for example, lack of funds for research, unqualified or overstressed faculty, deteriorating infrastructure, underuse of information and communications technologies (ICTs), stricter quality control and others. Moreover, universities are receiving less qualified students from the secondary level. The cited review shows that the performance of HEI can best be characterized by an environment where there exists policy instruments related to the development and finance of the traditional tasks of HEI, training, research and extension, but there is also indifference to research in the institutional context as well as resistance to changes in the institutional culture (Aguirre-Bastos and Rebois, 2011).

The above can be attributed to weak recognition of research in the normative architecture, which has a strong administrative bias, and the still limited existence of funding, human resources and equipment. Only more recently has there been better motivation of the research community and lines of research are being drawn based on national priorities. The weak recognition of research and extension produces discrimination or lack of incentives for active researchers, which are few in number and often self-made.

6.3.3 Science, Research, Technology and Innovation Policies

The national R&D policy (June 2010)

In April 1996 the government, through the then Ministry of Science, Technology and Higher Education, formulated its R&D policy, which was later revised by MCST (2008). The revision recognized that the previous policy had produced a limited number of achievements, and whose lack of implementation was associated with inadequate funding, resulting in a negligible S&T impact on the national economy.

At the same time, the driving force and preconditions behind the formulation of the 1996 policy were overtaken by many social and economic events in the sense that the plan did not address some critical issues related to the rapid technological change that had taken place at the end of the 1990s and beginning of the 2000s.

In this context, MCST formulated the 2010 plan, which later would give rise to and complement ongoing policy (MCST, 2010c) (see Aguirre-Bastos and Diyamett, 2015 for a discussion on the present policy). It should be considered, however, that the drafting of a new policy is only now being rethought after the completion of the national project on the review of the national system of innovation that took place in March 2015.[8]

MCST Master Plan

The Master Plan 2010–20 (MCST, 2010a) recognizes the country's weaknesses and limitations in STI and aims at setting out a long-term plan that will guide the implementation of the mandate of MCST in the next decade. The main aims of the Plan are:

a. to realign MCST with the current and future national priorities, immediate society needs, ongoing public sector reforms and stakeholders' and market demands, and
b. to accelerate the development of STI and harness its full potential and that of ICT to improve education and innovation, public sector management, private sector competitiveness in key strategic productive areas, industrialization, capacity building and poverty reduction in line with key sustainable development objectives as set in MKUKUTA, Vision 2025, the MDGs, as well as other national development policies and strategies.

The preparation of the Master Plan involved a large set of stakeholders in the public, academic and private sectors and civil society organizations. The Plan as described by Aguirre-Bastos and Diyamett (2015) recognizes the Ministry's long-term goal of becoming a catalyst that will enable the country to raise its productivity and competitiveness, and to improve the standard of living and quality of life of Tanzanians through the promotion and effective use of S&T, information and innovation processes, products and services in both productive and non-productive strategic sectors of the economy.

Sustainable Industries Development Policy 1996–2020

The Sustainable Industries Development Policy (MIT, 1996) sets out a first phase for consolidating existing capacities and a second phase (2000–10) for the creation of new capacities, focusing on areas with clear potential for gaining competitive advantages through the process of learning and application of efficient technologies. In this phase it is envisaged that new opportunities will be opened by taking into account emerging technological innovations. A third phase (2010–20) is envisaged as a full-fledged process of investments in basic capital goods industries.

The policy recognizes that one of the critical factors for sustained industrialization is the creativity and productivity of entrepreneurs, managers and workers, and that the present industrial sector is characterized by a relatively poor quality labour force, inadequate technical and managerial skills and a general lack of an entrepreneurial cadre exposed to advanced industrial culture. As a response, it is set out that government will develop the education system with due emphasis on technical training.

The policy also recognizes that domestic industry has inadequate research activities. The success of the sector is expected to be strongly based on scientific and technological research. Further, it is also recognized that there is a lack of linkages between the few local R&D institutions and the productive sector in the country, due to the fact that industrialists do not appreciate the role of R&D, and that R&D does not address the actual needs of the productive sector. The policy thus calls for the development of master plans, rationalization and synchronization of R&D institutions, articulation of areas for collaboration between manufacturers and R&D networks, particularly regarding financing and management of R&D institutions.

National Biotechnology Policy (June 2010)

The National Biotechnology Policy (MCST, 2010b) is envisioned to 'ensure that Tanzania has the capacity and capability to capture the proven benefits arising from health, agriculture industry and environmental applications of biotechnology while protecting and sustaining the safety of the community and the environment'.

The policy recognizes the challenges that its development represents and calls for the availability of human resources and funding. It also calls for the participation of the private sector in its development and the creation of a National Centre of Excellence for Biotechnology and Genetic Engineering that will create platforms to support the biotechnology industry.

The National Information and Communication Technology Policy (NICTP) (2006)

The NICTP policy, adopted in 2006, is set to facilitate the bridge of the 'digital divide' for Tanzania, thus transforming its economy into a knowledge-based economy, able to effectively compete in the global economy. This policy has now been revised but its main objective still remains the same. Its main results have already been mentioned in Section 6.2 of this chapter.

6.3.4 Research and Innovation[9]

In spite of the achievements that can be observed in the implementation of the different policies adopted throughout the years, these have been more the exception than the rule. There have been many limitations in reaching adequate R&D results, as recognized by MCST. Part of the wider problem has been the low level of inter-linkage and networking (Wangwe and Diyamett, 1998). Also, the rather weak and inadequate coordination that could have otherwise been offered and stimulated by a central national body such as COSTECH has often encountered a rather defiant stance by autonomous institutions under their own legal status.

One other key problem in the R&D system of Tanzania is related to the quality of the research that is undertaken. Low quality discourages policy makers and the private sector from using local research outputs and prompts them to seek research findings from abroad. To ensure research effectiveness, a performance monitoring and quality assurance system has to be in place. There exist neither institutional nor national mechanisms for assessing research performance by stakeholders in the country.

Following its several declarations, the government has put in place the National Fund for the Advancement of Science and Technology (NFAST). Nevertheless, inadequate funding has remained a major constraint to research and innovation as well as to education and training in STI. Constrained funding also explains the currently inadequately equipped R&D institutions, as well as outdated research facilities in tertiary institutions.

While there has been political will and commitment at the national level, there has not been a comprehensive management system to effectively plan, monitor and evaluate STI education, training and R&D activities in the country. Often, R&D activities are conducted in isolation and/or independently of a prioritized order of national and societal needs, with the result that research results – and any anticipated innovations – are known only to a few individual researchers.

There have been a number of potentially useful grassroots-level innovations reported by R&D institutions that have not been supported towards translation into tangible products, processes and services. There is clearly a need to diversify methods and strategies of disseminating research information and results beyond just research reports and journal articles. There is a need as well for regularized nationally supported ways and forums to broadcast results for purposes of systematic follow-up and/or informed decision making. Further, quality management of research in STI, partnership between R&D, industry and government and intellectual property rights (IPR) are some of the key aspects in a comprehensive management of STI that need to be introduced in the STI culture of the country.

It is often argued in Tanzania that research results are neither properly documented nor properly aligned to national and societal needs. The latter is a frequently cited problem in many developing countries, with which these authors do not fully agree, as research topics do follow national priorities. The problem for the use of research results must be sought for other reasons.

Patents (and other IPR forms) are a traditional source of measurement of the national capacity for technology development and innovation. There are two elements that need to be taken into account when dealing with patents as indicators to innovation. The first is the institutional setting as the national patent office is considered one of the institutions to be strengthened within the different S&T plans. The second element is the capacity of researchers and technologists to actually create patentable products. Until now this process has been very weak as evidenced from data from the patent office and other sources (see, for example, WIPO, 2010).

Finally, there are only limited efforts made to attract the private sector, individuals, business people, trade unions and community organizations into contributing significantly to the national STI effort by way of funding or shared sponsorship of research programmes. This is due in great part to the low level of technological capabilities of the productive sector.

In the social sciences there are several organizations of which the Economic and Social Research Foundation (ESRF), the Research for Policy Alleviation (REPOA) and the Science, Technology and Innovation Policy Research Organization (STIPRO) clearly stand out besides those existing at some of the larger universities, particularly the University of Dar es Salaam. In total, there are nine universities where social research is being undertaken.

ESRF represents an important and highly recognized organization in this area of knowledge. ESRF is an independent, non-governmental research centre founded in 1994 as a response to the need to develop capacity for policy analysis. An important activity of ESRF has been the launch in 2009 of the Tanzanian Knowledge Network (TAKNET) that provides a forum for exchanging and sharing information on social and economic development. STIPRO, on the other hand, is much newer but its studies are already well recognized as of high quality and providing a much needed information base to government decision-making processes. With this more general picture of the national system of innovation in mind, the sector systems of innovation in agriculture and health are analysed in some detail in the following sections.

6.4 THE AGRICULTURE SECTOR INNOVATION SYSTEM

6.4.1 Overview of the Agricultural Sector

As already discussed, agriculture is the backbone of the economy; 80 per cent of the country's population depends on it. Consequently, the country has directed many efforts to its strengthening and development. In more recent years, due to an increase in the budget allocated to agriculture (from 2.9 per cent in 2001/02 to 6.6 per cent in 2008/09), a positive trend has begun to emerge and the government has considered that this trend should be put into a structured frame. Such a framework is 'Kilimo Kwanza', meaning that the totality of the national development effort should be directed – on a priority basis – to the implementation of Tanzania's green revolution, as an ultimate vehicle for the social and economic transformation that the country requires.

There are several reasons why such a strategy is necessary, principally the fact that agriculture will continue to be for some time the backbone of the economy, food security is crucial, poverty reduction is a national objective, to reverse or at least attenuate rural/urban migration, use agriculture multiplier effects, contribute to alleviating health problems stemming from malnutrition.

6.4.2 Research Policy in Agriculture

According to the United Republic of Tanzania (URT, 2012b), the agriculture policy of 2012 asserts that the research system has continuously been reoriented to focus on a client-oriented approach by exploring practical and affordable options for engaging a broad range of individuals and institutions capable of contributing to sustainable livelihoods for the majority of Tanzanians. The major challenge is on how to work with a diverse spectrum of smallholder producers and entrepreneurs to develop farming systems that perform reliably and consistently in improving yield stability and safeguarding the productivity of land, labour and capital. The policy objective is to assure agricultural research services are strengthened in order to enhance productivity, competitiveness and profitability of the agricultural sector. In order to achieve the stated objective several policy statements have been put forward (URT, 2012b):

a. The agricultural research system shall be reformed to enhance the participation of a wide spectrum of stakeholders in identifying and setting research priorities.
b. A semi-autonomous institution shall be established to regulate and coordinate the agricultural research agenda.
c. Initiatives aimed at arresting agro-biodiversity deterioration shall be supported.
d. The government in collaboration with other stakeholders shall participate in funding and management of crop research.
e. The national research agenda shall be regulated and coordinated.
f. Indigenous knowledge shall be integrated into scientific research.
g. Public–private partnership in research activities shall be facilitated.
h. In collaboration with R&D institutions, research on irrigation and development of appropriate smallholder agricultural mechanization and agro-processing technologies shall be promoted; a research outreach arm for providing technology to farmers shall be developed.

i. The government shall conserve in a sustainable way the productivity potential of crop germplasm and related biodiversity in the existing agro-ecosystem such that it is not endangered by introduction of genetically engineered plants.

6.4.3 The Agriculture Sector Innovation System

The agricultural innovation system of Tanzania is comprised of several actors whose main functions are shown in Tables 6.1 and 6.2 (see Liwenga, 1988 for an early history of agricultural research in Tanzania).

Other actors include agro-industries, input suppliers, farmers, donors, cooperatives and financial institutions. Some groups of farmers, especially large-scale and commercial farmers, have been organized into cooperatives and associations, such as the Tanganyika Coffee Growers Association,

Table 6.1 Government STI agencies in agriculture

	Government agencies	Functions
1	Ministry of Communication, Science and Technology (MCST) www.mst.go.tz	Develop policies and guidelines and coordinate STI.
2	Ministry of Agriculture, Food and Cooperatives www.agriculture.go.tz	To deliver agricultural and cooperative services; build the cooperative and agricultural capacity of local government authorities; and to interface the government sector and public sector with regards to agricultural development.
3	Commission for Science and Technology (COSTECH) www.costech.or.tz	Fund, promote and coordinate research in Tanzania, including agricultural research, advises the government on research priorities.
4	Tanzania Food and Drug Authority (TFDA) www.tfda.or.tz	Standard certification, largely on food and drugs.
5	Tanzania Bureau of Standards (TBS) www.tbs.go.tz	Standard certification, general.

Source: Compiled from Larsen et al. (2009) and the individual institutions' web pages.

Table 6.2 Private and public research institutes and universities

	Institutes/universities	Functions
1	Sokoine University of Agriculture (SUA) www.suanet.ac.tz	SUA was established in 1984 with a primary focus on agriculture and veterinary sciences. SUA has four mandates: training, research, consultancy and outreach. The university offers undergraduate degrees in Biotechnology and Laboratory Sciences. Currently, the university has about 175 researchers trained at PhD level in all fields of agriculture, forestry and veterinary medicine. It has a Centre for Biotechnology, in collaboration with the International Centre for Genetic Engineering and Biotechnology (Trieste, Italy), with a major focus on agricultural biotechnology. The university has also established a technology transfer office to stimulate research translation.
2	Centre for Agriculture Mechanization and Rural Technology (CAMARTEC) www.madein-tanzania.com	The Centre carries out and promotes applied research to facilitate the design, adoption and development of machinery and equipment suitable for use in agricultural and rural development. It also develops and manufactures approved prototypes, components and cultural techniques and technologies and evaluates their suitability for adoption and alternative use in rural agricultural production. It performs tests on all types of machinery and equipment intended for use in agricultural and rural development and publishes the results of each test.
3	Tropical Pesticides and Research Institute (TPRI) www.tpri.or.tz	The Institute conducts research in tropical pests affecting plants, livestock and human health.
4	Tanzania Food and Nutrition Centre (TFNC) www.lishe.org	Carries out research and advises the government on all matters concerning food and nutrition.
5	Tanzania Fisheries Research Institute (TAFIRI) www.tafiri.or.tz	The Institute conducts training and research on fishery resources and disseminates research findings for fishery development and sustainable management of the resource.
6	Tea Research Institute of Tanzania (TRIT) www.trit.or.tz	Research on tea.

Table 6.2 (continued)

	Institutes/universities	Functions
7	Tanzania Coffee Research Institute (TACRI) www.tacri.org	Research on coffee.
8	Tsetse & Trypanosomiasis Research Institute (TTRI) www.fao.org/docrep/009/a0375e/a0375e00.htm	Research on animal health.
9	Directorate of Research and Development (DRD) www.erails.net/TZ/drd/drd-mafc/	Research on crops, livestock, natural resources, socio-economics.
10	Tanzania Wildlife Research Institute (TAWIRI) of the Ministry of Natural Resources and Tourism www.tawiri.or.tz/	Research on wildlife, especially preservation.
11	Tanzania Forestry Research Institute of the Ministry of Natural Resources and Tourism (TAFORI) www.taforitz.org/	Research and preservation of forests.

Source: Compiled from Yongolo and Mukama (2005) and the individual institutions' web pages.

cooperative unions in coffee-growing areas and tea and sisal growers associations.

Research activities in agriculture have picked up in terms of generating new seed varieties and obtained other isolated important results; however, the overall impacts of R&D projects are still minimal. Very few farmers, especially in the smallholder sector, have adopted the technologies developed by the public R&D centres. There are at least two major reasons for this; some of the new varieties meet productivity criteria but do not meet consumer taste and preference criteria because researchers paid too much attention to enhancing productivity and output and did not take into account consumer preferences, as well as inadequate dissemination of research findings with weak extension services.

In the sector, major decisions regarding research are made through the Directorate of Research and Development (DRD) in the Ministry of Agriculture and Food Security (MAFS). The Directorate also plans and executes research and disseminates research findings to those involved in agricultural production by providing extension services offered by another department within the Ministry. DRD operates through a network of research institutions, centres and substations that cover the main areas of crop research. There are seven zone research and training centres located in seven agro-ecological zones, which are responsible for applied and location-specific adaptive research.

Extension services in Tanzania have generally been understood as the transfer of agricultural technologies from experts (including progressive farmers) to farmers and other stakeholders such as livestock keepers. These services are used by DRD and its research institutions to transfer well-proven technologies from research institutions to farmers. The services have been the object of much debate and criticism among stakeholders in the agricultural sector. The weak links between research institutions and their partners, namely, extension services and farmers, have been widely recognized as one of the main reasons why the agricultural technology systems in Tanzania, and many countries in Africa, are ineffective and inefficient (Eponou, 1993; Expere and Idowu, 1990).

The sources of funds for R&D in agriculture include the central government, foreign donors, local government authorities, levies, contract research, revenue retention and cost recovery. Most R&D is being funded by external sources. About 80 per cent of expenditures made by R&D institutions come from donors. The government mostly covers salaries and other routine, non-developmental expenditures. In the past years, funding by both government and external donors has been reduced.

In 1994, the government launched two sources of agricultural research funds: the National Agricultural Research Fund (NARF) and the Zonal Agricultural Research Fund (ZARF). NARF has been used to finance research and links that cut across several agricultural zones. It has also been used to finance collaborative research that links government researchers and academic researchers. The Fund has demanded high-quality research proposals from applicants, but only a limited amount of resources have actually been disbursed. NARF has not been built as fast as had been anticipated because there has been a failure to attract additional research funding from other sources. Lengthy review procedures for proposals have contributed to the delay.

ZARF has been used to fund zone-specific research as part of decentralization, empowerment of local stakeholders to fund and influence research and a move towards financial sustainability. District councils and local stakeholders are encouraged to contribute to this Fund.

Starting in 1994, the agricultural research institutes were allowed by the government to retain income generated from the sales of produce and services, user fees, rentals and consultancy contracts. The amounts collected increased from TAS 59 million in 1994–95, to TAS 271 million in 1999–2000, and then declined somewhat to TAS 185 million in 2001–02.[10]

To promote the use of agricultural inputs, the government has established the National Agricultural Inputs Trust Fund to provide loans through local banks. The results of different analyses attempting to relate the relevance of research to sources of funding show no significant correlation between successful or unsuccessful projects and sources of funding. An ample assessment of the evolution of the Tanzania agricultural sector from the perspective of investments, policy and regulations has recently been conducted by the Alliance for a Green Revolution in Africa (AGRA, 2014), showing the advances made in the sector but also the slow speed of the different processes leading to an effective improvement of its operation.

Public–private partnerships are beginning to emerge in the landscape of R&D institutional arrangements in agriculture. Collaboration between the Agricultural Research Institute (ARI), Mlingano (a public R&D organization) and Katani Ltd (a private organization owning sisal estates and sisal processing factories) is a public–private partnership. The project, which was funded by donors, took an innovative approach. Instead of the research funds being managed by ARI Mlingano, they are being managed by Katani Ltd, the main beneficiary of the research outputs. The proposal for the research grant was jointly prepared by Katani Ltd and ARI Mlingano. The evidence suggests that under such arrangements the R&D system is more responsive to the needs of the farmers compared with the situation in which government funds are managed by public research institutions.

Two experiences stand out in the private sector initiative as coffee and tea growers associations now own and manage coffee and tea research institutes. The Tea Research Institute (TRIT, 2004) has a very efficient system for identifying priority areas, conducting R&D and disseminating information to end users. Since privatization in 1996, annual output has increased from 15,000 tons to 28,000 tons – an increase attributed to research work. What emerges from existing analysis is that farmers, even the smallholder peasant farmers, are willing to contribute to research activities when they have been impressed by the outcomes of research. The successful privatization of tea research has been replicated for coffee and cashew nuts.

6.5 THE HEALTH SECTOR INNOVATION SYSTEM

6.5.1 Overview of the Health Sector

Investments in education and health in the recent past have enabled Tanzania to record some improvement in the Human Development Index (HDI) ranking, from position 163 in the pre-MKUKUTA period to 151 in 2009, falling again in 2013 to 152 among 187 countries and 159 in 2014 among the same 187 countries (UNDP, 2014).[11] In this way, it has moved from the low human development group of countries to the mid human development group of countries.

Life expectancy has increased to 55 years (54 years for males and 56 years for females), mainly due to declines in HIV and AIDS prevalence and child mortality. The neonatal mortality rate has improved only marginally partly due to little improvements in maternal health.

Improvement in infant and under-five mortality rates have been largely due to gains in malaria control through improved diagnosis and treatment of malaria, as well as prevention through increased use of insecticide treated nets. In addition, success of other preventative measures, such as measles vaccination, vitamin A supplementation and implementation of Integrated Management of Childhood Illnesses (IMCI) have enhanced child survival. However, disparities persist between and within regions and districts, urban and rural areas, and by wealth status.

As noted in the review of the health system by USAID (2013), the country faces several challenges that impede reaching national development goals; such challenges include the shortage of health care workers, health commodity stock-outs and insufficient financing (West-Slevin and Dutta, 2015). Although medical schools and enrolment have increased over the years, the shortage of skilled health professionals, which is most felt in rural areas, is estimated to be 65 per cent. Moreover, existing health professionals are unevenly distributed, with significant disparities within regions and within districts. These and other weaknesses have been recognized by government in its review of the sector strategic plan (URT, 2013b).

6.5.2 Policy and Actors in the Health Sector Innovation System

The National Research and Development Policy of 2010 recognizes the importance of conducting research and the utilization of evidence in the improvement of health services and to guide the health sector research in the country.

The Tanzanian health research priority setting process is led by the National Institute for Medical Research (NIMR, 2002, 2008) and health research prioritization is regarded as a key factor in an effort to strengthen the national health research system. It is expected that the product of priority setting exercises would provide a direction to researchers, policy makers and other stakeholders to effectively identify research areas that are likely to make an impact and lead to improved public health outcome. Additionally, prioritization mechanisms are necessary to meet the current demand for increased harmonization of health research at a global level to realize the MDGs related to health, that is, MDG number 4, 5 and 6 (as discussed in Chapter 2).

In funding, the Health Research User's Trust Fund (HRUTF) was formed in 1996 by NIMR to enhance demand-driven health research efforts. HRUTF addresses health research problems identified by stakeholders. Established with seed money of $177,300 from the Swiss Development Cooperation, it is supported by an annual subvention of TZ50 million from the Ministry of Health.

In the health sector there are several organizations led by NIMR that report to the Ministry of Health. NIMR is both a clearing house for medical research and a coordinating agency for resource mobilization for R&D activities in the health sector. In addition, it provides a link between research institutions and the government. The Institute also oversees research activities at three centres and three research stations as well as research activities at teaching hospitals.

There are some inherent limitations with such an arrangement. The most common complaint from beneficiaries is the lack of involvement by medical professionals working with hospitals. It is felt that medical R&D lacks an appropriate dissemination forum or feedback mechanism (in spite of the existence of the TANHER Forum[12]) that could increase the usefulness and effectiveness of medical research.

Faced with this limitation, NIMR has embarked on several new approaches to disseminate its research results; in addition to a health research bulletin, annual reports, annual joint scientific conferences and radio and television programmes have been identified as effective communication channels. Regular media workshops are held to bring together both print and broadcast media to discuss the results of medical research.

Decades of research on a wide range of diseases like malaria, filariasis, plague, schistosomiasis and more recently on HIV/AIDS has been undertaken, and there are other numerous diseases that could be the subject of interesting research. Many researchers are of the opinion that malaria and HIV/AIDS are priority areas for medical research in Tanzania that warrant additional resources (TWAS, 2009). As will be discussed further, researchers believe their work is of use to both policy making and applications in the health system, in particular, in the improvement of health care quality and better patient management.

Table 6.3 Government STI agencies in health

	Agencies	Functions
1	Ministry of Communication, Science and Technology (MCST)	Develops policies and guidelines and coordinates STI.
2	Commission for Science and Technology (COSTECH)	Promotes, funds and coordinates research, and advises the government on research priorities.
3	Ministry of Health and Social Welfare	Provides health policy guidance and regulation of the system.
4	Medical Stores Department	The main procurer, supplier and distributor of essential medicines for the public sector, and the primary supplier to faith-based organizations and non-commercial groups providing health services.
5	Tanzania Food and Drug Authority (TFDA)	The country's key regulatory body responsible for controlling the quality, safety and effectiveness of food, drugs and medical devices.
6	Tanzania Bureau of Standards (TBS)	Standard certification of products.

Source: Compiled from Shah et al. (2010).

Tables 6.3, 6.4 and 6.5 identify and describe the functions of the main actors in the health sector system of innovation; the tables indicate that Tanzania has a rich and complex health innovation landscape, with the public sector driving the innovation agenda.

6.6 'CLOSING THE LOOP' IN AGRICULTURE AND HEALTH: EMPIRICAL EVIDENCE

6.6.1 Objective

Against the above background on the two sector systems of innovation, two case studies were devised. The overall objective of the two cases was to access empirical evidence on how science-based research and innovation is designed, implemented and applied in Tanzania in support of development and poverty alleviation – in line with the closing the loop concept.

Specifically, the study aimed at (a) determining the social and economic

Table 6.4 Research institutes and universities

	Institutes/universities	Functions
1	National Institute for Medical Research (NIMR)	Plays a critical role in medical research in Tanzania. Its current research priority areas are malaria, filariasis, trypanosomiasis, onchocerciasis, schistosomiasis and sexually transmitted diseases including HIV/AIDS. NIMR has around 87 full-time research scientists and 34 technical staff.
2	University of Dar es Salaam (UDSM)	The university has a biotechnology unit within its Faculty of Science, where annually around 35 students are trained at the undergraduate level and 10 students at the graduate level. Health research includes the identification of antimicrobial agents from Tanzania's biodiversity.
3	Muhimbili University of Health and Allied Sciences (MUHAS)	A strong medical research centre, with strong expertise in traditional medicine, clinical trials and diagnostics, and with strong overseas partnerships. MUHAS houses the Institute of Traditional Medicine, which plays an important role in traditional medicine research in the country. The institute has a herbarium with an inventory of over 3000 plants being used locally in traditional medicine. It is working in the areas of diabetes, benign prostatic hypertrophy and identification of antimicrobial agents. It has a pilot manufacturing plant that has been producing herbal preparations including mixtures for peptic ulcers, cough, asthma and cholesterol since 2003.
3	Sokoine University of Agriculture (SUA)	A Centre for Biotechnology, in collaboration with the International Centre for Genetic Engineering and Biotechnology (Trieste, Italy). SUA collaborates with NIMR on research on cross-cutting issues between health and agriculture.
4	Ifakara Health Institute	Best known for research on malaria, but has also expanded to other areas including tubeculosis, HIV and other communicable diseases.
	Kilimanjaro Christian Medical College (KCMC)	KCMC has developed capabilities in health biotechnology and clinical trials.

Source: Compiled from Shah et al. (2010).

Table 6.5 Private sector

	Private sectors	Functions
1	Tanzania's pharmaceutical industry	Manufacturing of medicines
2	Zenufa	Manufacturing of medicines
3	Keko Pharmaceutical Industries	Manufacturing of medicines
4	Interchem Pharmaceuticals	Manufacturing of medicines
5	Shelys (Dar es Salaam)	Manufacturing of medicines
6	A to Z Textile Mills (Arusha)	Manufacturing of insecticide impregnated bed nets

Source: Compiled from Shah et al. (2010).

relevance of the knowledge that is being generated in research institutions; (b) identifying sources of funding for knowledge generation and dissemination; (c) determining the status of dissemination and use of knowledge; and (d) identifying challenges facing researchers in the production and dissemination of knowledge.

6.6.2 Research Method

Study areas and population

The target population included universities, R&D institutions and policy makers. The universities and R&D organizations included for the agricultural sector were Sokoine University of Agriculture (SUA) and Mikocheni Agricultural Research Institute (MARI), and for the health sector, Muhimbili University of Health and Allied Sciences (MUHAS) and the National Medical Research Institute (NIMR).

Sample size and sampling procedure

A simple random sampling was used to obtain a sample size of 30 researchers from each sector, representing the above organizations. In agriculture, all researchers surveyed responded. However, for the health sector only 16 researchers successfully filled out the questionnaire. In addition, a purposive sampling was used to get six key informants: four as head of the institutions and two policy makers for the two sectors.

Data collection

The major data collection methods for the primary data were questionnaire and an interview guide. In addition, existing information (secondary information) was collected to enrich the primary data.

Table 6.6 Level of education

Research area	Master's degree		PhD	
	Frequency	Per cent	Frequency	Per cent
Agriculture	14	46.7	16	53.3
Health	2	12.5	14	87.5

Source: Survey data, 2013.

Data analysis

Descriptive analysis was used to analyse quantitative data while content analysis was used to analyse qualitative data. Quantitative data collected from researchers were summarized, coded and entered into the software program, Statistical Package for Social Sciences (SPSS), for analysis. To provide the quantitative description of information, frequencies and percentages were obtained and used to present results.

6.6.3 Research Findings

Basic information about the respondents

Table 6.6 provides a general overview of the respondents' educational level. In agriculture, 46.7 per cent of the respondents hold a Master's degree and 53.3 per cent hold a PhD degree. For researchers in health sciences, 12.5 per cent had a Master's degree and 87.5 per cent hold a PhD degree. Thus, the majority of researchers in both health and agriculture research institutions have PhDs. This can be attributed to capacity development initiatives at research institutions whereby researchers possessing Master's degree are encouraged to pursue PhD training and accorded opportunities through various projects implemented in these institutions.

Further, the sample was heavily biased towards male researchers, accounting for almost 85 per cent of the respondents, somewhat higher than the average rate of 75 per cent for male researchers in all scientific areas in Tanzania (see Table 3.1). The age distribution of the respondents shows that half of the researchers were 45 years or older, 35 per cent between 31 and 45 years old, and 15 per cent 30 years or under.

Social and economic relevance of the research and quality of research

Social and economic relevance of research can be measured by looking at how the research reflects the needs of the people who are the end users of the research results. In general, research that is directed towards solving problems affecting large groups of people is considered useful to society.

Table 6.7 Major purpose of the research carried out in the institution

Research area	Generation of products: number. and (%)	To inform public policy	Both generation of products and to inform public policy
Agriculture	7 (23.3)	8 (26.7)	15 (50.0)
Health	0 (0.0)	10 (62.5)	6 (37.5)

Note: Numbers in parentheses are percentages.

Source: Survey data, 2013.

Recognizing the importance of linking research to all walks of social and economic development of people, and in the spirit of closing the loop concept, the two case studies solicited information on the major purpose of the research carried out in the sample institutions, as shown in Table 6.7. Note that there are two major utility values of research: to generate new products and processes and to inform policy.

According to the findings in Table 6.7, research in both agriculture and health is carried out for the two major purposes described above: to inform policy and for commercial and non-commercial (goods and services) direct use by end users. The respondents were asked to indicate one of the three options as the focus of their research: to inform policy; for products and services; and both policy and products and services. While researchers in the agricultural sector seem to focus on both of the two utility values equally, that is, almost 50 per cent for both generation of products (goods and services) and to inform public policy, the health research is tilted more towards informing public policy. This could arise from the fact that even new products and processes, such as drugs required in curing diseases like malaria as well as new vaccinations to prevent disease, influence policy decisions for the health sector. For instance, it is the government who decides which type of drugs should be used for combating certain diseases. This is probably due to a predisposition of the health sector towards being more of a public sector agent, while agriculture is half way between private and public.

Participation of policy makers and end users in the research team

One of the important objectives of the study is to find out the extent to which research in the two sectors is relevant for the socio-economic development of Tanzania; and this to a large extent concerns issues of research agenda setting. The issues include how the priorities were determined; whether or not they have included policy makers and other

Table 6.8 Participation of policy makers and end users in the research team

Research area	Yes		No	
	Frequency	Per cent	Frequency	Per cent
Agriculture	13	43.3	17	56.7
Health	6	37.5	10	62.5

Source: Survey data, 2013.

end users during the research project's formulation, including taking part in the research team. Table 6.8 indicates the extent of participation of end users of research in agenda setting as well as being part of the research team.

The findings in Table 6.8 show that the participation of end users in the research project formulation is below 50 per cent for both sectors. This finding indicates that there is low participation of the end users in the research agenda setting. This situation can be attributed to the nature of the formulation of research projects whereby end users are only thought of as recipients of the research results, a fact that seems to be responsible for the low uptake of research in the country.

Most important factors considered during research project formulation

There are normally a number of major driving forces that trigger the formulation of a research project, or even the thinking of initiating one. Four major factors seem to be behind most research formulation in poor developing countries at the level of an individual researcher. These are: compliance with national development goals; following the agenda of the organization; following the priorities of funding agencies; and possibility publishing in peer-reviewed journals (academic impact). The two case studies also sought to establish the forces behind the formulation of research projects in Tanzania; the researchers were asked to mention one or more of the above influences as reasons for initiating research projects. The result is shown in Table 6.9.

The data in Table 6.9 indicate that the choice of project topic is very much influenced by funding agencies. Almost half of the respondents in both agriculture and health asserted that research priorities are determined by the priorities set by funding agencies. In-depth interviews with key informants further revealed that this is because many organizations do not have a budget specifically put aside for research; most research is based on research project proposal writing, where in most cases areas of focus have been predetermined by the funding agencies. For some research projects, researchers have been able to define the research themes following the national development plans or their institutional research agenda, but these are very few. Most of the organizations have their research priorities and plans in place, but these are rarely followed because of the reasons just explained above. In addition, several researchers active in agricultural sciences at universities indicate that for them academic excellence is the determining factor. This is especially true for research leading to academic degrees. As one of the university-based informants put it: 'for students and their supervisors, the research topics are largely determined by what is possible and academically exciting'.

Funding of research

Having a good idea is only one part of the research process. Having the funds to carry out the scheduled work is another matter. The study attempted to investigate the accessibility of sources of funding for research as well as the opinion of the researchers on the balance of funding between research and dissemination of research results (Tables 6.10 and 6.11).

Sources for research financing in Tanzania include government funding, donor funding, funds from the private sector and own funds. According to the results of the multiple responses, 73.3 per cent of the respondents from agriculture indicated that they had received government funding, 90 per cent had at some time received donor funding, 23.3 per cent had access to funding from the private sector and, perhaps surprisingly, 13.3 per cent used own fund in one or more of their research projects. The corresponding percentages for the respondents from health institutions were 100 per cent for government funding, 100 per cent for donor funding and 43.8 per cent funding from the private sector. These findings are in line with the fact that most research projects for health are equally funded by both donors and government. For agriculture, the responses show that donors are a more important funder of research than the government. However, in terms of size, and across all research carried out in Tanzania, the contribution of the donors is almost half (Madikizela, 2013).

The private sector seems to be another source of funding to some researchers. Private sectors such as non-governmental organizations (NGOs) have a tendency to fund researchers who are conducting their research in line with the institutions' goals and objectives. However, few researchers are able to fund small research projects from internal sources as some reported that they are able to conduct their research projects using funds generated from sales of produce or services such as consultancies.

Table 6.9 Factors considered by researcher as most important when formulating research projects

Research area	Compliance with national development goals		Compliance with faculty/ department research priorities		Compliance with priorities of funding agencies		Academic impact	
	Frequency	Per cent	Frequency	Per cent	Frequency	Per cent	Frequency	Per cent
Agriculture	12	27.3	4	9.1	21	47.7	7	15.9
Health	9	31.0	6	20.7	14	48.3	0	0

Source: Survey data, 2013.

Table 6.10 Sources of funding available for research (multiple responses)

Research area	Government funding		Donor funding		Fund from private sector		Own fund	
	Frequency	Per cent	Frequency	Per cent	Frequency	Per cent	Frequency	Per cent
Agriculture n = 30	22	73.3	27	90.0	7	23.3	4	13.3
Health n = 16	16	100.0	16	100.0	7	43.8	0	0.0

Source: Survey data, 2013.

Table 6.11 Underfunding in the process of research

Research area	Knowledge generation (research)		Knowledge dissemination		Both	
	Frequency	Per cent	Frequency	Per cent	Frequency	Per cent
Agriculture	9	30.0	16	53.3	5	16.7
Health	6	37.5	9	56.3	1	6.2

Source: Survey data, 2013.

The 'closing the loop' concept involves two major stages of research: knowledge generation and knowledge dissemination. Availability of funds is equally important for the two stages, although in practice, the dissemination stage is normally overlooked. With regard to funding, the survey sought to investigate which one of the two is usually underfunded in the case of Tanzania. Results show that 30 per cent of respondents from agriculture opined that knowledge generation is underfunded, 53.3 per cent said knowledge dissemination and 16.7 per cent said both. In health, 37.5 per cent of the respondents said knowledge generation, 56.3 per cent said knowledge dissemination and 6.2 per cent said both (Table 6.11). The findings confirm the general tendency of overlooking the dissemination stage in the research process. Some researchers expressed the opinion that the problem of knowledge dissemination always depends on who has funded the project and whether or not the project is purely for academic purpose. Some funding agencies put emphasis on dissemination and therefore the activity is included in the project planning phase. Otherwise, according to the majority of respondents, dissemination of research results is something that is usually not planned for during the initial stages of research.

Extent to which donor funding of research is aligned to national development goals

As observed above, most of the research projects in Tanzania are being funded by donors, and there is a general understanding that donor funding is bad because it does not focus on national development goals. This study also sought to verify this contention by seeking the opinion of the researchers themselves. The responses are indicated in Table 6.12.

As seen in Table 6.12, there is a big difference between the health and agricultural sectors. While a good number (75 per cent) of all researchers in the health sector were of the opinion that donors do abide by the requirements of the national development goals in making funding decisions, only

Table 6.12 Adherence of funding agencies to the alignment of research objectives with national development goals

Research area	Yes		No	
	Frequency	Per cent	Frequency	Per cent
Agriculture	13	43.3	17	56.7
Health	12	75.0	4	25.0

Source: Survey data, 2013.

about 43.3 per cent of agricultural researchers said that they do. One can assume from these findings that health research has much more impact on the lives of people than agricultural research because of a better match between funding and relevance. This is probably because – although some health problems are specific to localities – generally health is one of the major global issues and donor priorities are on issues that are global. We would imagine even greater disparities between research funding and priorities in the industrial sector.

Responding to an additional question that aimed at understanding whether the current government decision to increase funds for research would solve the above problem, most said this is not a panacea, because there is no national research agenda that guides priorities for research in different sectors. Many organizations (universities and R&D organizations) try to draw their own research priority from national development plans and visions. However, they think that there should be a national competent body that draws a national research agenda from which other organizations can draw their plans. This was especially emphasized by the respondents from universities. Discussions with officials from COSTECH, who to a large extent are responsible for national research priority setting, indicate that COSTECH is currently reviewing its national priorities for research in different sectors, and the priorities will be made available to research organizations in the near future.

Dissemination of research results

A most important issue in the 'closing the loop' concept is dissemination of research outputs. An active dissemination strategy for research products/results is important if research outputs are to reach intended users. Researchers were therefore asked to give their opinion on the opportunities they have in sharing their research results with policy makers and end users in the process of closing the loop. The results are indicated in Table 6.13.

Table 6.13 Opportunities that exist for researchers to share research results with policy makers and end users

Research area	Always		In rare cases		Not at all	
	Frequency	Per cent	Frequency	Per cent	Frequency	Per cent
Agriculture	5	16.7	25	83.3	0	0.0
Health	7	43.8	9	56.2	0	0.0

Source: Survey data, 2013.

The data in Table 6.13 indicate that a majority (83.3 per cent and 56 per cent) of researchers in both agricultural and health research institutions feel that they have rare chances of sharing their research findings with policy makers and end users. A majority of researchers do not have forums where they can interact with decision makers; instead, they share their findings with colleagues prior to publication. It was also indicated that because policy makers have limited time to read detailed research reports, researchers often disseminate information to them in the form of policy briefs, brochures and executive summaries that highlight actionable recommendations for decision making. Furthermore, during the in-depth interviews researchers argued that policy makers do not have time to read even those short summaries. However, the respondents acknowledged the fact that if the research problem is defined in collaboration with potential users of research outcomes, and properly negotiated with them, the chances for use very much increase.

Extent of use of research results

Logically, next to relevance of research priorities is the extent to which the research has been put to use. Most researchers in both the health and agricultural sectors feel that their research results have satisfactorily been put to practical use. Table 6.14 indicates the opinion of researchers on the extent of use of research results while Table 6.15 shows the perceived gap of collaboration between scientists and policy makers.

The findings of Table 6.14 show that over 70 per cent of the respondents in both agriculture and health assert that their research outcomes have either satisfactorily or to a great deal been put to use. This may seem contradictory with the findings in Table 6.15 on the linkage between researchers and users where 93.3 per cent and 93.8 per cent of respondents from both the agricultural and health sectors, respectively, said that there is a poor linkage between researchers and users of results, which corroborate

Table 6.14 Are research results being put to use?

Research area	A great deal		Satisfactorily		Rarely		Not at all	
	Frequency	Per cent	Frequency	Per cent	Frequency	Per cent	Frequency	Per cent
Agriculture	4	13.3	19	63.3	1	3.3	6	20.0
Health	6	37.5	8	50.0	2	12.5	0	0.0

Source: Survey data, 2013.

Table 6.15 Is there a gap of collaboration between the scientific community and policy makers and other end users of research?

Research area	Yes		No	
	Frequency	Per cent	Frequency	Per cent
Agriculture	28	93.3	2	6.7
Health	15	93.8	1	6.2

Source: Survey data, 2013.

findings of other researchers (see, for instance, Coulson and Diyamett, 2012; Shah et al., 2010).

Because of this rather unexpected response, a follow up using in-depth interviews with selected respondents was made. Most of the respondents in this category feel that the use of research results is on the lower side than expressed by the surveyed researchers because of poor linkage between researchers and users of research outputs. They further explained that most of the organizations have technology transfer offices, which are supposed to collect potential research outputs from researchers from the faculty. However, since the transfer activity does not count in assessments and promotions, nobody really cares about communicating their research outputs to the technology transfer office; neither do the staff of these offices aggressively search for potential research outputs for transfer activities.

Socio-economic impact of research: study findings and some success stories

In addition to the question on the extent to which research outputs have been put to use, researchers were also asked about the socio-economic impacts of their research outputs. The responses are indicated in Table 6.16. Empirical findings from the case studies were also complemented by existing success stories as presented in this section.

The study indicates that most of the researchers were of the opinion that their research projects had impacts on development in one way or the other. About 92 per cent of the health researchers feel that their projects had an impact on policy, 57 per cent on social development and 28 per cent on economic development (Table 6.16). These findings are in line with the above responses on the main purpose of research where respondents in the health sector responded that their research is to a large extent policy oriented (see Table 6.7). This can be explained by the fact that even for product development and use, the sector is highly regulated – it is government who decides what type of drug

Table 6.16 Research impact (multiple responses)

Research area	On policy		On social development		On economic development	
	Frequency	Per cent	Frequency	Per cent	Frequency	Per cent
Agriculture n = 22	11	50.0	18	81.8	17	77.3
Health n = 14	13	92.9	8	57.1	4	28.6

Source: Survey data, 2013.

should be used for different ailments. According to existing information, the pharmaceutical industry in Tanzania does not use knowledge generated in Tanzania to manufacture drugs. This is partly because of the low capabilities of the pharmaceutical sector in Tanzania, but also due to IPR issues. According to Shah et al. (2010), researchers at NIMR had developed an innovative extraction process for the active ingredient of Artemisia annua, a plant from which artemisin compounds for the treatment of malaria is extracted. However, according to Shah, the compound has never been commercially exploited because of IPR problems: scientists were reluctant to share the information outside the institution because of lack of intellectual property protection. Shah et al. (2010) also point out that the Chemistry Department at University of Dar es Salaam has a research group focused on medicinal chemistry and traditional medicine. According to Shah et al. (2010), this group has established linkages with GIBEX, an international consortium on biotechnology research with a focus on biodiversity, bio-exploration and natural product research including identification of wild mushrooms as a source of biomedical products.

The above general picture notwithstanding, there are some isolated success stories as indicated in Shah et al. (2010). Some of these stories include those of researchers at MUHAS. In the 1990s, for instance, Professor Zul Premji, a researcher at MUHAS, played a critical role in the development of the rapid diagnostic test for malaria and its validation along with the company Beckton Dickinson. This rapid diagnostic test is now widely used and available as a product produced by various local and international companies. Professor Fred Mhalu is another pioneer researcher at MUHAS. His research focuses on adapting point of care diagnostics to local settings for various infectious diseases such as malaria, HIV/AIDS and tuberculosis. According to Shah et al. (2010), Professor Mahalu completed the phase I study of a multigene, multi-clade HIV

DNA candidate vaccine in Tanzania as part of a long-standing Swedish-Tanzanian collaboration.

In regard to the agricultural sector, about 50.0 per cent of the researchers said that their projects had an impact on policy, 81.8 per cent on social development and 77.3 per cent on economic development. However, these high numbers would appear to contradict the results of previous studies done by other researchers who found that agricultural research in Tanzania has not been successfully connected to agriculture (see, for example, Coulson and Diyamett, 2012). According to Coulson and Diyamett, the reason behind this failure is an unstructured technology transfer system. Other reasons identified included lack of information centres to serve farmers and extension agents as well as low demand from farmers for research-related services. Moreover, interactions between farmers and extension agents for most cases were found to be very weak, resulting in a low rate of adoption of research results. In other areas, farmers reported that the technologies are not appropriate, not available, not known, too complex, too expensive, not profitable, and too risky or not convincing (Isinika, 2012).

The above weaknesses notwithstanding, like the health sector, there are examples of isolated success stories. For instance, the Uyole model has been credited for a mini-green revolution in the Southern Highlands. A report by Isinika (2012) indicates that there was a strong research team that worked closely with a strong on-station social economic department. These teams interacted regularly with extension agents and farmers through field days to evaluate technologies. There were annual meetings between researchers and extension agents in each of the Southern Highlands regions. Researchers were accountable to the regions for answers to problems raised during the previous year. The system was complemented by improved road and railway infrastructure and inputs subsidy.

Another success story comes from the effective application of client-oriented research in the Northern zone of Tanzania where farmers and extension agents participate in the research process through on-farm trials and farmers' field schools (Isinika, 2012). Yet another lesson comes from a pigeon peas study. Pigeon peas are an important food crop in certain parts of Tanzania, including the region of Babati. However, field studies carried out in 2004 in Babati showed that in many areas of Tanzania production of pigeon peas was declining due to a plant fungus called *fusarium wilt* (Bekele et al., 2007, cited in Coulson and Diyamett, 2012). To respond to this, in the early 1990s, the research stations and ICRISAT in Nairobi started breeding varieties that would resist this disease, and also mature more quickly. Improved varieties were released in 1999, 2002 and 2003.

These varieties possessed good aromas and colour, and could be cooked easily. The study showed that the farmers who adopted the new varieties were substantially better off. Following this, efforts to multiply the seeds and promote the new varieties were intensified at Selian and Ilonga research stations. By 2010, researchers from Selian reported that the adoption rate had risen to 80 per cent, with yields sharply rising. They also commented that much of the seeds had been multiplied on the research stations or on growers' farms under the supervision of the research stations. These successes were a result of joint efforts between the government of Tanzania and local and international R&D institutions.

Challenges facing researchers in carrying out research

According to the study, researchers in Tanzania are faced with the following challenges: limited funds from government and donors; poor motivation among researchers in terms of working conditions; limited opportunity of collaboration with other researchers in the region/subregion; lack of knowledge transfer departments, and an absence of good policy to support such programmes; inadequate provision of IPR; a persistent gap in collaboration between researchers, policy makers and end users of the research results.

In response to these challenges, many of the respondents suggested that there should be a clear strategy for knowledge transfer. This should start with the way research projects are conceived. In this respect, the stage of knowledge generation is the most problematic. To align knowledge generation to the social and economic needs of the country, there has to be a clear national research agenda developed sector-wise; and there should be adequate budget attached to it. To ensure uptake of generated knowledge, research projects must be negotiated with the users during the design stage. Most importantly, presentation of results must be done carefully to minimize confrontation. Some respondents think that there is not a good relationship between government ministries and universities because of the way universities present results in the name of academic freedom.

6.7 CONCLUSIONS AND RECOMMENDATIONS

6.7.1 Policies and the Innovation System

Tanzania, like all developing countries, faces significant internal and external challenges in the new global economy that affect not only the shape and mode of operation but also the purpose of their STI and higher education systems. Among the most critical dimensions of change are the convergent impacts of globalization, the increasing importance of knowledge as the main driver of growth and the information and communications revolution.

STI and higher education are in fact the crucial factors for creating and maintaining sustainable growth. The existing long-term poverty reduction goals and other national objectives can only be achieved if the focus on capacity building of STI and higher education is enhanced. In fact, economic growth is increasingly tied to the supply of individuals with advanced research-based education in S&T and with ample experience in research.

In spite of large efforts made by the government and donors in S&T development, these have yet to make an important impact (with some exceptions) to a change of economic and social conditions. Recent economic growth has been brought about mainly by favourable external conditions, while at the same time there is persistent underdevelopment and poverty conditions, as demonstrated by the country's difficulties in attaining the MDGs (as discussed in Chapter 2).

Under such circumstances it is difficult to visualize the vision set to change the country's productive status into a semi-industrialized country by 2025. Despite being rich in infrastructure, its national system of innovation seems to be in disarray: although there are isolated best practices, and successful cases, in general the way in which science-based research and innovation is designed, implemented and applied in Tanzania does not seem to take Tanzania to where it wants to go.

There is observed a weak recognition of research in the normative architecture, which has a strong administrative bias, and the still limited existence of funding, human resources and equipment. Only more recently has there been better motivation of the research community and lines of research are being drawn based on national priorities. The weak recognition of research and extension produces discrimination or lack of incentives for active researchers, which are few in number and many of which are self-made.

Analysing all policy documents and strategies so far defined, it is recognized that Tanzania requires a roadmap that is realistic and clearly identifies what can really be done in the short and long term. Extensive and overoptimistic lists of objectives and goals, in overlapping policies, weak coordination mechanisms, little funding, a disarticulated innovation system, weaknesses in human resources and other limitations cannot lead the country into an effective capacity building process.

Pending further analysis, the R&D policy is too ample and ambitious to be implemented in the short or medium term. It contains excessive details, but at the same time it does not point out how several objectives and goals will be reached. It simply calls for new norms, mechanisms, institutions to be put in place. It would seem of little use to have both an STI Policy and Plan and an R&D Policy and Plan. Further there is also the MCST Master Plan and COSTECH Plan. These should be brought under a single framework.

In general, it is clear that in spite of the efforts made by the government to enforce the policies it formulates, there remains a problem of coordination among various sectors. Often, the different ministries, when formulating their own sector policies, have either ignored or just glossed over the central policy issues that could constitute a common agenda. In this way, these 'sector' ministries have found themselves either inadvertently in opposite directions against each other or giving preference for internal ministerial interests. The need for tighter and binding coordination of policy cannot be overemphasized.

6.7.2 Closing the Loop

Considering the results of the survey conducted under this study, there seem to be inconsistencies in researchers' responses in regard to different issues addressed in this project. For instance, most researchers have said that their research outputs to a large extent have been put to use; but at the same time argue that linkage between them and users of research outputs, such as policy makers, is extremely limited. It is difficult to explain such contrasting differences, but one explanation could be that there are other ways that research outcomes reach the intended users, other than communication by researchers themselves – there can be unintended spillovers.

But most likely it can also be that researchers do respond that their research is being put to use, and having impacts, even if they know this is not the case. Researchers are concerned about funding for research, and they know that knowledge use will most likely attract more funding. However, the second scenario is more probable if one compares the findings of this study with other existing studies that indicate that a large part of the knowledge generated is not put to use. On the other hand, the difference in response between the health and agricultural sector is close to reality; that is, health sector research is more likely to be put to use than agricultural sector research because of its public sector nature and therefore close follow-up by the government.

A major conclusion to be drawn here is that evidence-based public policies are very important in closing the loop, whether in the public or private sectors. This general conclusion notwithstanding, however, there is a need to carry out a comprehensive survey of users of research to countercheck responses by researchers. It is also important to include other sectors. One observation of this study is that there exists a big difference between the health and agricultural sectors; where health is very much dominated by policy-oriented research, while agriculture is much less so. It would be interesting to compare this with, for instance, the industrial sector, which is mostly run by private enterprises.

The government must decide whether or not to have an efficient and top level think tank in the public sector, well linked to private think tanks or rely on the latter for policy making. The improvement of the policy and planning context will directly impact STI and higher education policies. The country requires knowledge-based decision-making processes. Scientifically informed decisions are key to improving governance.

6.7.3 Agriculture

There are extensive efforts in R&D in the agricultural sector conducted by both public and private organizations. In spite of this, R&D projects have had minimal impacts on agriculture, which is dominated by many smallholder farmers, especially those producing food crops. This group was concerned, for example, that new varieties did not satisfy consumer preferences and although productivity had increased, failure in the market led to discontinuation of these new varieties.

R&D policy definition for this sector must be a joint effort between government, universities, farmers and other stakeholders

In the rural environment, farmers should come together in order to create productive associations, and then initiate a process of creating 'collective trademarks', 'collective certificates of origin', receive support (for example, from universities) on marketing, quality assurance and others to overcome existing bottlenecks for the introduction of their products into the local and export markets. There are already interesting international experiences on this type of 'collective' entrepreneurship.

6.7.4 Health

Tanzania's medical R&D is relevant and timely as it addresses issues of national importance. Medical research focuses on endemic diseases that affect rural areas, where up to 80 per cent of the population earn their living. Malaria is the most prevalent and serious vector-borne disease. More work has been done on malaria than on any other disease. Relevance is achieved because the main owner of R&D activity is also the formulator of health policy, and the major diseases are location specific.

The effectiveness of medical R&D is reduced by poor communication. As a result, information does not reach the beneficiaries in a way that is effective in solving problems. Additional effort is needed to involve stakeholders and better inform them of research results.

NOTES

1. BuddeCom, www.budde.com.au (accessed 30 June 2015).
2. www.indexmundi.com/tanzania/#economy (accessed 30 June 2015).
3. Worldbank.org/data (accessed 30 June 2015).
4. See also MPEE (2006) for an evaluation on the progress of the attainment of the Millennium Development Goals.
5. Approximately USD 14 million at the present exchange rate (16 August 2015),
6. This document provides a complete view of the education sector, trends and mid-term objectives, including the educational budgets up to 2017.
7. http://www.worldbank.org/indicator (accessed 7 July 2015).
8. Results still need to be drawn and analysed.
9. Much of the information presented here was gathered from Wangwe et al. (2009).
10. No other sources could be found for later years.
11. Rankings in such a long period are not strictly comparable as the methodology for their determination has changed in the past three years.
12. The Tanzania National Health Research Forum (TANHER) was established in 1999 as 'a consultative and advisory body to policy and decision makers as regarding health research coordination, undertaking, collaboration, dissemination of health research results, and enhancing utilization of research results for policy and decision-making'. The initial grant of USD 200,000 was provided by the Rockefeller Foundation.

REFERENCES

AGRA (2014), 'An assessment of agricultural policy and regulatory constraints to agribusiness investment in Burkina Faso, Ethiopia, Ghana, Nigeria and Tanzania', Alliance for a Green Revolution in Africa (AGRA), Nairobi, Kenya.

Aguirre-Bastos, C. and R.R. Rebois (2011), *Review and Evaluation of the Performance of Tanzania's Higher Education Institutions in Science, Technology and Innovation*, Final report to UNESCO, Dar es Salaam, 23 June.

Coulson, A. and B. Diyamett (2012), 'Improving the contribution of agricultural research to economic growth: policy implications of a scoping study in Tanzania', International Growth Centre (IGC) Working Paper.

Diyamett, B., P. Ngowi and M. Mutambala (2011), *The Role of Foreign Direct Investment (FDI) in Local Technological Capability Building: The Case of Tanzania*, African Technology Policy Studies (ATPS), Tanzania, May, available at http://www.stipro.or.tz/index.php?option=com_jdownloads&task=do (accessed 23 June 2016).

Eponou, T. (1993), *Partnership in Agricultural Technology Generation: Linking Research and Technology Transfer to Serve Farmers*, Research Report No. 1, The Hague: International Service for National Agricultural Research (ISNAR).

Expere, J. and I. Idowu (1990), 'Managing the links between research and technology transfer: the case of agricultural extension research liaison services in Nigeria', ISNAR Linkages Discussion Paper No. 6, International Service for National Agricultural Research (ISNAR), The Hague.

Isinika, A. (2012), 'The relationship between research and extension in Tanzania: the role of district councils', available at http://www.theigc.org/sites/default/files/research_extension_linkage.pdf (accessed 18 May 2013).

Larsen, K., R. Kim and F. Theus (eds) (2009), *Agribusiness and Innovation Systems in Africa*, Washington, DC: World Bank.

Liwenga, J.M. (1988), 'History of agricultural research in Tanzania', in J.M. Ter and A.Z. Mattee (eds), *Proceedings of a Workshop on Science and Farmers in Tanzania*, Morogoro, Tanzania: Sokoine University of Agriculture (SUA), cited by B.L.M. Mwamila and B.D. Diyamett, 'The evolving role of universities in economic development', in B. Göransson and C. Brundenius (eds), *Universities in Transition*, Springer, pp. 171–90.

Madikizela, M. (2013), *Mapping Research Systems in Developing Countries, Country Report: The Science and Technology System of Tanzania*, Paris: UNESCO.

MCST (2008), The National Science, Technology and Innovation Policy (draft), Ministry of Communication, Science and Technology, Dar es Salaam, 1 April.

MCST (2010a), Science and Technology and Communication Master Plan 2010–20, Section I – Science, Technology and Innovation, Preliminary Summary prepared by G.R. Bamwenda, Dar es Salaam, May.

MCST (2010b), National Biotechnology Policy, Ministry of Communication, Science and Technology, Dar es Salaam.

MCST (2010c), The National Research and Development Policy, Ministry of Communication, Science and Technology, Dar es Salaam, June.

MIT (1996), Sustainable Industries Development Policy (SIDP) (1996–2020), Ministry of Industry and Trade, Dar es Salaam October.

MOEVT (2010), *Higher Education Development Programme 2010–15*, Ministry of Education and Vocational Training, Dar es Salaam, March.

MPEE (2006), *Millennium Development Goals: Progress Report*, Ministry of Planning, Economy and Empowerment, Dar es Salaam, December.

Mutambala, M., J. Liberio and B. Diyamett (2013), 'Research project on institutional technology transfer policies and strategies in East Africa: Lessons from universities and R&D institutions – a background paper for Tanzania'. STIPRO, Dar es Salaam.

NIMR (2002), *Annual Report*, Dar es Salaam: National Institute for Medical Research.

NIMR (2008), *The Fourth Tanzania National Health Research Priorities 2013–2018*, Tanzania National Institute for Medical Research.

Shah, R., P.A. Singer and A.S. Daar (2010), 'Science-based health innovation in Tanzania: bednets and a base for invention', *BMC International Health and Human Rights*, **10**(Suppl. 1), S4.

TAAS (Tanzania Academy of Sciences) (2007), 'TAAS news', **1**(1), December.

TRIT (2004), *Annual Report, 2002/03*, Tea Research Institute of Tanzania.

TWAS (2009), *Ifakara Health Institute, Excellence in Science: Profiles of Research Institutions in Developing Countries*, Trieste: Academy of Sciences for the Developing World.

UNCTAD (2002), *Investment Policy Review of the United Republic of Tanzania*, New York and Geneva: United Nations Conference on Trade and Development.
UNCTAD (2013), *World Investment Report*, New York and Geneva: United Nations Conference on Trade and Development.
UNDP (2014), *Tanzania Human Development Report 2014: Economic Transformation for Human Development*, New York: UNDP.
UNIDO (2012), *Tanzania Industrial Competitiveness Report*, Dar es Salaa:m UNIDO and Ministry of Industry and Trade of the United Republic of Tanzania.
URT (2000), *National Development Vision 2025*, Dar es Salaam, United Republic of Tanzania.
URT (2005), *National Strategy for Growth and Reduction of Poverty*, Vice President's Office, Dar es Salaam, United Republic of Tanzania, June.
URT (2006), *Medium Term Strategic Plan: 2006/07–2009/10*, Ministry of Higher Education, Science and Technology, Dar es Salaam, United Republic of Tanzania.
URT (2008), *Higher Education Development Programme 2010–15*, available at http://www.tanzania.go.tz/egov_uploads/documents/Higher_Education_Development_Programme.
URT (2010), *National Strategy for Growth and Reduction of Poverty (NSGRP II)*, Ministry of Finance and Economic Affairs, Dar es Salaam, United Republic of Tanzania.
URT (2011), *Integrated Industrial Development Strategy 2025*, Ministry of Industry and Trade, Dar es Salaam, United Republic of Tanzania.
URT (2012a), The Tanzania Five Year Development Plan 2011/12–2015/16, *Unleashing Tanzania's Latent Growth Potentials*, Dar es Salaam, United Republic of Tanzania.
URT (2012b), Ministry of Agriculture Food Security and Cooperatives National Agriculture Policy, Dar es Salaam, United Republic of Tanzania.
URT (2013a), *State of the Tanzania Economy*, Report prepared for the National Poverty Policy Week.
URT (2013b), Mid-term Review of the Health Sector Strategic Plan III 2009–2015, Dar es Salaam, United Republic of Tanzania.
USAID (2013), 'USAID/Tanzania Health System Strengthening Strategy 2013–2018', 30 September.
Wangwe, S.M. and B.D. Diyamett (1998), 'Cooperation between R&D institutions and enterprises: the case of URT', in *Atas XI Bulletin, Approaches to Science and Technology Cooperation and Capacity Building*, New York and Geneva: United Nations.
Wangwe, S., B. Diyamett and A. Komba (2009), 'Trends in research and development in Tanzania: funding sources, institutional arrangements, and relevance', in M. Graham and J. Woo (eds), *Fueling Economic Growth: The Role of Public–Private Sector Research in Development*, Ottawa: IDRC and Practical Action Publishing, pp. 159–92.
West-Slevin, K. and A. Dutta (2015), *Prospects for Sustainable Health Financing in Tanzania*, Washington, DC: Health Policy Project, Futures Group.
WIPO (2010), *World Intellectual Property Indicators*, Geneva: World Intellectual Property Organization.
World Bank (2009), *Tanzania Country Brief*, Washington, DC: World Bank, May.

World Bank (2015a), *Tanzania Overview*, Dar es Salaam, available at http://www.worldbank.org/en/country/tanzania/overview.
World Bank (2015b), *Tanzania Mainland Poverty Assessment*, Dar es Salaam, available at http://www.worldbank.org/tanzania.
Yongolo, C. and B. Mukama (2005), *Development of S&T System and Experience of Tanzania on S&T Data*, Dar es Salaam, Commission for Science and Technology.

7. The national innovation system in Mozambique and its relevance for development

Carlos Aguirre-Bastos, Sérgio Chicumbe, Maximiano Dgedge and Bo Göransson*

7.1 INTRODUCTION

In 1975 Mozambique attained its independence from Portugal, but it was only in 1992 with the establishment of a peace agreement after the extended civil war that the country was able to initiate its development process including the undertaking of research and development (R&D) efforts. Government and international cooperation provided the support for the R&D programmes and projects, mainly in public institutions and universities, as well as for disseminating its outputs.

Research outputs are certainly needed for informing policy and transferring them to accelerate economic and social developments. In this context, the project 'Closing the Loop' seeks to ensure the awareness, understanding and ownership of research outputs principally but not exclusively by decision makers. Its goals are to increase the relevance and utilization of research outputs, thereby enhancing the influence of researchers, organizations and institutions on the development process.

The project is expected to contribute to a better understanding of how the process of knowledge creation and innovation can be made more conducive to the development efforts being pursued in Bolivia, Mozambique, Tanzania and Vietnam. Particular emphasis has been given in the analysis of the agriculture and health sectors. This chapter describes the results of the project in Mozambique.

Section 7.2 of the chapter provides an overview of the present national political, social and economic context, discussing in particular the observed economic growth and its impact on social inclusion and the fight against poverty, principally in the rural areas. Section 7.3 describes the national science and technology (S&T) policy and the national innovation system.

Moreover, it provides the latest collected indicators that allow the measurement of the performance of the system.

Section 7.4 briefly describes the present situation of agriculture in the country as well as the existing policies and research outputs in the agricultural sector innovation system. It assesses the performance of the system through an extensive literature review of policy documents as well as policy analyses that have been undertaken in past years, and which serve as the main basis for this section.

Section 7.5 describes the main characteristics of the health sector and assesses the policies and research outputs in the health sector innovation system, again through a literature review of policy documents and policy analyses of the performance of the sector innovation system.

Section 7.6 presents a survey conducted among researchers active in the areas of agricultural sciences and health sciences. In total, the opinions of 63 researchers on the quality and impact of research in Mozambique were obtained through a consultation process with the scientific community. Finally, concluding remarks on the state of the national and sectoral innovation systems are given in Section 7.7. It closes with a discussion on the present efforts of the Mozambique government to better define the science, technology and innovation (STI) policy and strategy and to improve sector outputs as well as their utilization.

7.2. POLITICAL, SOCIAL AND ECONOMIC CONTEXT

Mozambique is still one of the world's poorest countries. Economic mismanagement, a brutal civil war from 1977 to 1992 combined with floods and droughts led the country to severe economic crisis. To overcome this situation, the country adopted a market economy policy after reaching an agreement with the Bretton Woods institutions in 1987. The government embarked then on a series of macroeconomic reforms designed to stabilize the economy. These steps, combined with heavy donor assistance and political stability generated after the 1992 peace agreements, and the multi-party elections in 1994, have led to dramatic improvements in the country's growth rate.

The gross domestic product (GDP) has been growing continuously since 2000, making the country one of the fastest and most stable growing economies in Africa. In 2005 the GDP was USD 6.5 billion and in 2013 it had gone up to USD 15.5 billion at current prices. The growth rate in 2014 stood at 7.4 per cent and it is forecasted that this rate will continue up to 2017.[1] The World Bank attributes this growth to sound macroeconomic management, large-scale foreign direct investment (FDI) projects, political stability and significant donor support.

In the specific case of investment, the net FDI reached USD 6.7 billion in 2013, up from USD 1.3 billion in 2011, representing a growth rate of 29 per cent in the period 2007–13. Most of the FDI focused on megaprojects in the extractive sector, specifically coal, gas and aluminium, which are the main products for export. These megaprojects are, however, capital intensive and do not generate enough jobs to provide sufficient opportunities for the fast growing young population or contribute towards poverty reduction.

Looking into the future, the World Bank notes that the recent discoveries of large deposits of coal and gas may transform Mozambique into a significant player in global markets. However, it is also noted that to reap the benefits of a growing resource sector the country will need to develop the capacity to manage the extractive industries and ensure that they contribute to sustainable and broad-based growth.

The impact of donor support to development has been amply documented in recent years. Pivodic and Nhachengo (2014) discussed a new concept of absorption capacity when dealing with large inflows of external funds, especially for research. Their analysis in the health sector of Mozambique shows that it is necessary to advance the concept of absorption capacity by taking a people-centred approach that acknowledges that health systems are shaped by how leaders and managers exercise authority.

It is argued by these authors that the traditional thinking of absorption capacity neglects important aspects of the funding relationship and needs to move the focus away from absorption towards the more holistic notion of the 'quality' of a funding relationship. This altered perspective entails shifting attention towards those individuals who conceive and administer a funding relationship and it transfers some of the responsibility for the effectiveness of aid towards the 'donor' and thus creates a more equitable share of responsibility for the success of funding. It is argued that advancing the absorption capacity paradigm towards a focus on relationships between individuals involved in the funding relationship can increase the effectiveness of external funding of health systems.

Official Development Aid (ODA) in Mozambique was USD 2.3 billion in 2013, an increase from USD 1.2 billion in 2004.[2] In relation to foreign aid, Torvinen (2013) discusses the issue of governance, concluding that the country is still unable to take its own decisions due to the external influences it suffers from donors. This and other assessments make very little reference to issues directly related to the governance of science, research and innovation.

An evaluation of budget support operations undertaken from 2005 to

2012, made by ITAD, FISCUS and MBConsulting (2014) and commissioned by the Evaluation Unit of the European Commission, states that budget support has effectively strengthened the structure for policy dialogue based upon the definition of policy targets. It also points out that the main structural weakness of the policy dialogue process lies within the policy development process, which should be situated at the sector and thematic levels. Some critical issues are recognized that need to be dealt with: productivity in agriculture, high levels of malnutrition and a poor enabling environment for doing business, where effective policy responses have not been generated. Although the evaluation does not deal with S&T directly, it is clear that the above observations require the notice of S&T policy makers.

On the social front, it is generally considered that the rapid economic growth of past decades has not been translated into significant poverty reduction and principally in rural poverty that remain almost unchanged in many areas of the country. This view is debated by James et al. (2005) who argue that observed growth is only pro poor if it reduces inequality. In their vision the observed pattern of growth in Mozambique is definitely pro poor, as inequality between provinces and regions has declined over time, although growing inequality in Maputo City is a cause for alarm and policies must be adopted to counteract this trend.

Measurement of the changes in social conditions can be observed from the achievements of the Millennium Development Goals as shown in Table 7.1 for the period 1997–2013.

In spite of the observed improvements, existing indicators show in general an extremely weak social situation; today the country ranks 178th out of 187 countries in the Human Development Index (HDI) in 2014 (UNDP, 2014).

7.3 S&T POLICIES AND THE NATIONAL INNOVATION SYSTEM

7.3.1 STI: Institutional Developments and Policy

Considering the importance of S&T for development, Mozambique created in January 2000 the Ministry of Higher Education, Science and Technology. Later in 2005, through Decree 13/2005, the Ministry of Science and Technology was established and in January 2015 it became the Ministry of Science, Technology, Higher and Technical Education. It is expected that continued institutional change will not affect the policy direction that the country has undertaken since 2008.

In recognition of the role that S&T plays in the fight against poverty

Table 7.1 The Millennium Development Goals (MDGs) for Mozambique

Goals	Indicator	1997	2003	2007	2008	2009	2015	Situation of the supportive environment	State at a glance – will the target be reached??
1. To eradicate extreme poverty and hunger	Population living below the National line (%)	69.4	54.1	n/d	n/d	54.7	40	Strong	Potentially
2. To achieve universal primary education	Enrolment rate in primary school (%)	n/a	n/a	n/d	81.3	n/d	100	Reasonable	Potentially
	Completion rate of primary school (%)	22.0	38.7	n/d	77.1	n/d	100		
	Literacy rate (%)	39.5	46.4	n/d	46.9	n/d	n/a		
3. To promote gender equality and empower women	Ratio of girls to boys in primary school (%)	0.71	0.83	n/d	0,9	n/d	1.0	Strong	Potentially
	Ratio of literate females to males (as % of ages 15–24)	0.62	0.83	n/d	n/a	n/d	1.0		
4. To reduce child mortality	Under-five mortality rate (per 1000 live births)	245.5	154	147.2	n/d	138	108	Strong	Potentially
	Rate of infant mortality (0–1 year, per 1000 live births	143.7	101	95.5	n/d	93	63		

5. To improve maternal health	Maternal mortality rate (per 100,000 live births)	692.0	n/a	500	n/a	n/a	250	Reasonable	Potentially
6. Combat HIV/ AIDS, malaria and other diseases	Rate of prevalence of HIV/ AIDS among adults (15–49 years of age)	8.6	9.3	11.5	n/d	11.5	n/a	Reasonable	Potentially
6. Ensure sustainability of the environment	Proportion of land area covered by forest	21.9	n/a	51	n/a	n/a	n/a	Reasonable	Potentially
8. Develop a global partnership for development	Net development aid received as percentage of GDP	n/a	25.2	n/a	18.3	n/a	n/a	Reasonable	Potentially

Source: National Institute of Statistics (INE).

and in the equitable and sustainable economic growth of the country, combined with the absence of a legal framework in the sector at the time, the Council of Ministers adopted in 2003 the S&T Policy through resolution No. 23/2003 of 22 July.

This policy provides the main guidelines that the country is to follow; its established Mission and Vision are given respectively as '[T]the policy should contribute to the reduction of absolute poverty, to improve the standard of living of the populations and to national development, through the creation of conditions to foster research, encourage and promote scientific and technological competences, including local knowledge, and through the promotion of more effective processes of technology transfer and endogenous technology production' and '[T]he S&T policy is a basic policy, infrastructural and transversal governance. It is a policy of innovation, based on the exchange between the national and international scientific capacity and the utilization of comparative advantages. Promotes innovation and scientific and technological development integrated, complete and competitive, ensuring excellence in strategic thematic areas chosen in function of their potential contribution to economic growth, the alleviation of poverty and improvement of living conditions.'

It is important to note that since the first policy statements in Mozambique, the concept of innovation has been readily introduced. This situation is quite different from other developing countries where the innovation element of policy was brought at a much later date, after the S&T components were more developed.

The 2003 policy had four specific objectives under the main goal of aiming to develop an integrated system of production and knowledge management facing national needs in order to boost the sustainable development of the country:

a. Develop an integrated research and innovation system.
b. Strengthen the national education system and professional training in the component of scientific and technological capabilities.
c. Develop the innovative capacity of the national productive sector.
d. Develop a system of dissemination and communication related to scientific and technological knowledge, taking advantage of the new information and communication technologies.

In compliance with the adopted policy, the government defined on 27 July 2008 the Mozambique Science, Technology and Innovation Strategy (MOSTIS) as a planning tool with a time horizon of ten years. MOSTIS contains seven key strategic objectives (Nhamusso, 2013), aligned with the strategic and priority actions included in the policy.

a. Foster a culture of innovation throughout Mozambican society.
b. Promote S&T and innovation within the industrial and public sectors and promote the transfer of technology.
c. Promote the use of information and communications technologies (ICT) for good governance and service delivery, and for the diffusion of knowledge, in support of poverty reduction and economic growth.
d. Promote human resource development at all levels in the areas of STI.
e. Build and improve the policy instruments, institutions and infrastructure of the S&T system.
f. Establish funding policies and mechanisms for research and innovation.
g. Promote the mainstreaming of S&T within all sectors.

7.3.2 The Mozambique Innovation System: Actors and Functions

Since the adoption of MOSTIS, a number of initiatives and actions have been taken in order to establish an operating dynamic Mozambique innovation system (MIS) as required by policy. Such efforts have resulted in the creation of legislation, regulation and institutional developments.

Of particular importance were the restructuring of the Ministry of Science and Technology, Regional Centers of Science and Technology (CRCTS), the National Research Fund (FNI), Thematic Scientific Councils, the Academy of Science of Mozambique (ACM), the National Institute of ICT (INTIC), S&T parks and others. There were also approved regulations and laws such as electronic transactions law, status and mobility of researchers, research licensing and registration of research institutions. Some of the main actors and functions are briefly discussed later in this section.

The MIS comprises a complex heterogeneous network of institutions and organizations, public and private, as represented in Figure 7.1. Some of the more important research institutions are included in Table 7.2, and a brief discussion on each of the key levels follows.

7.3.3 Policy Management Level

The level of S&T policy management is through the Ministry of Science, Technology, Higher and Technical Education (MCT) and different line ministries and public institutions that develop, promote and coordinate the implementation of the S&T and sector policies. The institutions are responsible for creating favourable conditions for the operation of the MIS, namely, through a system of incentives, collaboration and mobility.

The Mozambican government comprises 28 ministries whose tasks are to formulate and implement sectoral policies, strategies and regulations

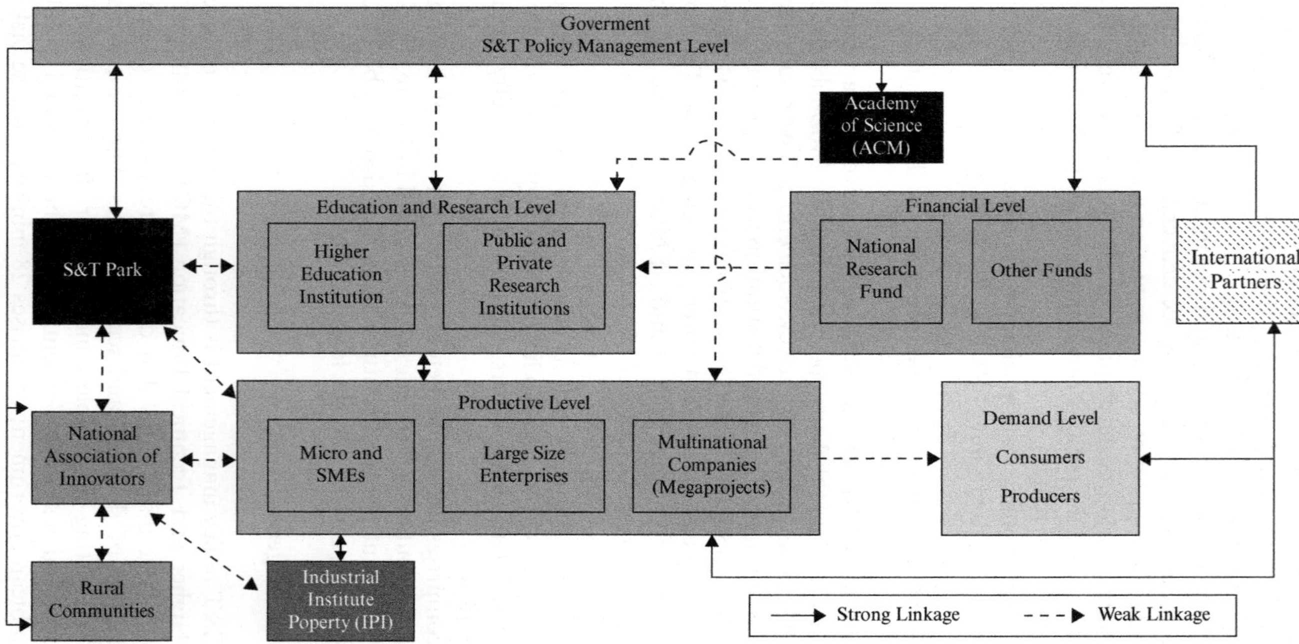

Source: Ministry of Science and Technology, February 2015.

Figure 7.1 The Mozambique innovation system (MIS)

Table 7.2 Research institutions in the Mozambique national innovation system

Institution	STI units	Main activity
Ministry of Transport and Communication	National Hydrograph and Navigation Institute	Applied research, monitoring and extension
	National Meteorology Institute	Applied research and monitoring
Ministry of Health	National Health Institute	Basic and applied research
	Health Research Centre at Manica	Basic and applied research, monitoring and extension
	Regional Health Development Centre	Monitoring and extension
Ministry of Agriculture	Agriculture Research Institute	Applied research, monitoring,extension and development programmes
	National Cartography and Tele-detection Centre	
	Institute for Animal Protection	Animal production
	National Institute for Agronomy Research	Research
	National Institute for Veterinary Research	Research and extension
	Centre for Forestry Experimentation	Research and extension
Ministry of Fisheries	Institute for the Development of Small Scale Fishing	Extension and development programme
	Institute for Fisheries Research	Applied research and monitoring
Ministry of Public Works & Housing	Engineering Laboratory of Mozambique	Applied research
Council of Ministers	National Statistics Institute	Applied research
Ministry of Education and Culture	National Institute of Educational Development	Applied research
	Institute of Socio-cultural Research	Applied research
	Eduardo Mondlane University (see further discussion on universities in this study)	Basic and applied research and extension

Table 7.2 (continued)

Institution	STI units	Main activity
	Pedagogical University	Basic and applied research and extension
	Higher Institute of International Relations	Applied research
Ministry of Industry and Commerce	National Institute for Normalization and Quality	Development of norms, standards and quality control
Ministry of Justice	Centre for Juridical and Judicial Training	Training
Ministry of Culture	Institute for Sociological and Cultural Research	Anthropological culture and history

Source: Mouton (2007).

as well as their monitoring and evaluation. Theoretically, these central organs have to interact with the different actors of the MIS, particuly at the education and research level as well as the productive level with the purpose of increasing productivity and national income. The different line ministries not only monitor and evaluate the performance of S&T policy as referred to their sector, but also allocate financial resources for research programmes and mobilize resources and create incentive systems to stimulate research and innovation programmes.

MCT is the central body with the mission to design and coordinate the implementation of the STI policy in Mozambique alongside sector ministries. In this role, MCT carries out inspections of all actors that perform research activities in order to ensure scientific rigour, ethics and quality. It also has a central role in funding STI through the National Fund for Research (NRF), as described later. The Ministry also has an important operational role in the management of international cooperation agreements in the context of STI.

The success of implementation of the S&T policy depends on the internal process of planning at MCT, particularly the articulation among the central organ,[3] the S&T Provincial Delegations and District Technology Services (SDEJTs) (Figure 7.2).

It is considered in this chapter that the internal planning process has been deficient and marked by weak articulation among the activity plans designed by the central organ, Provincial Delegations and District Services.[4] The activity plans seem to be disconnected or unrelated due to very weak coordination and communication among organic units from

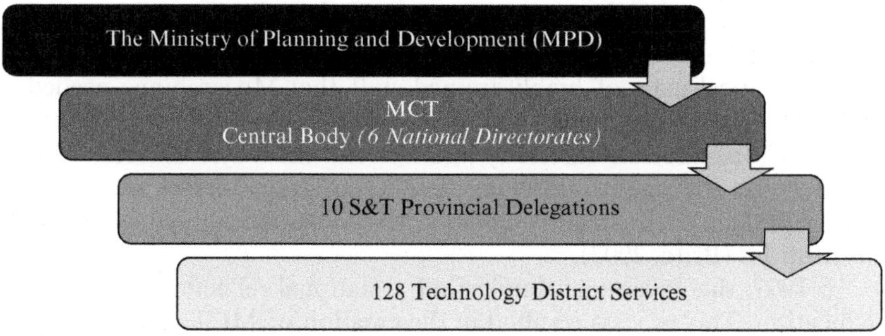

Source: Council of Ministers, MOSTIS, June 2006.

Figure 7.2 The internal planning process in MCT

MCT. Thus, the central organ does not have capacity to influence the provincial and district plans.

Once S&T is a cross-cutting issue, the implementation of the S&T policy becomes a complex process and requires the involvement of different ministries. Strong collaboration between the institutions is crucial to the successful implementation of MOSTIS and consequently the S&T policy.

Evidence shows, however, that there are problems of coordination among MCT, other ministries and public institutions. The role of MCT is often misinterpreted and arouses competition or conflicts of interests. There are instances when line ministries do not collaborate with MCT in the implementation of the STI programmes and projects.

In general, the system concentrates excessively on sector plans to render evaluations to government and parliament and the sector plans do not include coordination processes at the expense of cross-cutting issues and poor communication.

7.3.4 The Education and Research Level

Higher education in Mozambique has been considered a priority sector since independence. The first higher education institution in Mozambique was created in 1962 by the Portuguese as a branch of the Portuguese universities with the primary aim of educating the settlers' children in Mozambique (Chilundo, 2002). This institution was called Estudos Gerais Universitários. It was later upgraded to the University of Lourenço Marques in 1968. This first university discriminated against indigenous Mozambicans. In fact, until 1974 indigenous Mozambican students constituted less than 0.1 per cent of the student population (Brito, 2003).

At independence this university was transformed into Eduardo Mondlane University (UEM) in 1976 with the main mission of training the critical mass of highly trained staff that Mozambique needed to solidify the independence of the new Republic. Therefore, UEM became the first Mozambican university and it is still the major university in the country. As the nation's foremost public university, UEM has played a leading role in the reform processes in the Mozambican higher education institutions (Brito, 2003).

In 1997, the government undertook a national reflection about higher education. This process resulted in the creation of MCT in January 2000 and the approval, in August of that year, of the *Strategic Plan of Higher Education in Mozambique 2000–2010*. The approval of this plan was followed by the preparation of an operational plan that divided the ten-year implementation period into three phases: (1) the reform phase (2000 to 2004); (2) the expansion phase (2005 to 2007); and (3) the consolidation phase (2008 to 2010). The changes that occurred in higher education in this period emerged in the course of this transformation process.

In the mid 1980s, two more public universities were created, the Higher Pedagogical Institute (ISP) and the International Relations Institute (ISRI). Since then the number of private and public higher education institutions has been growing rapidly. By 2000 there were eight higher education institutions of which five were private; by 2011 this number had grown to 38 institutions of which 21were private and 17 public. The evolution in the number of higher education systems may be observed in Figure 7.3.

Statistical data from the Ministry of Education show that the number of graduates in different fields of knowledge was 3500 in 2000 and increased

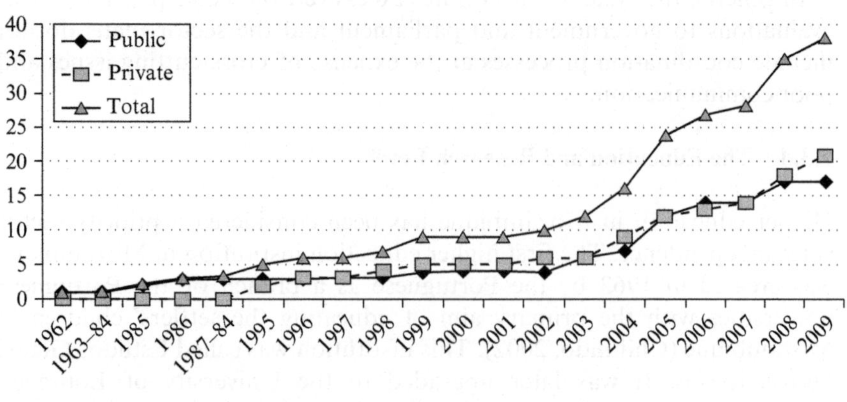

Source: Ministry of Education, 2013.

Figure 7.3 Evolution of institutions of higher education, 1975–2009

to 4000 in 2013, representing a growth of around 14 per cent. However, most students remain unemployed due to low demand in the labour market; some studies claim that education curricula do not fit the needs of the productive sector, and is the main reason for the high demand of foreign labour.

The second task of universities is to develop R&D activities to create a stock of knowledge. The results of the national R&D survey carried out in 2012 showed that 30 per cent of higher education institutions developed some kind of R&D activities and they have appropriate infrastructure. It also found that financial resources were invested more in applied research (57 per cent) than in basic research (25 per cent) and experimental development (18 per cent).

One other task of the university is connected to business. Modern universities are not only interested in the creation of knowledge but also its commercialization, particularly in the establishment of companies, for example, in the way of 'spin-offs', S&T parks and incubators. In Mozambique, evidence shows that higher education institutions concentrate on training human resources and they do not perform the business task and establish few partnerships with micro, small and medium-sized enterprises (SMEs), large and multinational companies. However, the Mozambican government has made efforts to overcome this scenario through the establishment of S&T parks and incubators.

R&D activities are undertaken by higher education institutions and also public and private research institutions. Evidence shows that there are few private research institutions and that there is poor collaboration between them and with public institutions mainly due to limited human and financial resources as well as the weak influence of policy measures. The R&D survey results identified 30 private companies that had R&D activities during 2010. The small number of private research institutions is partly responsible for the high propensity to import and adopt technologies from abroad.

The education sector is in general an important performer of R&D and a source of innovation in the country. Most of the university research is carried out by public universities, specifically UEM, Universidade Pedagógica (UP), Universidade Unilúrio, Universidade UniZambeze, Instituto Superior Politécnico de Gaza (ISPG) and Instituto Superior Politécnico de Manica (ISPM). All of these public universities run under the Ministry of Education. Some private universities such as Universidade Católica, Universidade São Tomás, Universidade Técnica de Moçambique (UDM), Instituto Superior Técnico de Moçambique (ISQTEM), Universidade Mussa Bin Bique and others have also been engaged in research activities in the past few years.

UEM has the largest research volume with about 300 ongoing projects at present. Almost all research activities been funded by external organizations (MOSTIS, 2006).

7.3.5 The Financial Level

Funding of R&D and innovation is done through three main channels:

a. The State budget that is assigned to the different ministries, public research institutions and public laboratories by the Ministry of Finance.
b. Financial resources placed on a competitive basis by the FNI, which private and public institutions and individual researchers can access.
c. Local and foreign financial resources channelled to higher education institutions, private research institutions and companies under the form of grants, credit and capital transfers.

The results of the third national S&T survey show that financial resources were applied in a total of 1456.6 million Meticais[5] in R&D programmes during 2010. From the total amount, 66.9 per cent comes from donations, 22.3 per cent from the State budget, 10.5 per cent from other resources and 0.3 per cent credit. It is clearly seen that donor funding is the most important source. The major source of external funding used on R&D in the period 2008–10 came from international partners (World Bank, Swedish International Development Agency (SIDA), Norwegian Agency for Development Cooperation (NORAD), European Union, Finland and Japan).

When scientific publications are analysed it is found that in the last eight years or so, more than 90 per cent is produced in association with an external partner[6] (SCImago Lab, 2015), mainly based on donor funding.

Figure 7.4 provides a picture of the funding situation. It can be seen that expenditure on R&D as a percentage of GDP (GERD) has increased in the period 2008–10 due partly to government efforts in mobilizing resources for research programmes. In 2008, GERD stood at 0.38 per cent of gross national product (GNP) while in 2010 it increased to 0.46 per cent and further increased to 0.61 per cent in 2012 (MST, 2014). It is noted that these values are still below 1 per cent as suggested by NEPAD.

With the establishment of the FNI in 2005 an important door for access to funding by all stakeholders in the innovation system through competition was established (Mouton, 2007). FNI provides resources to applicants

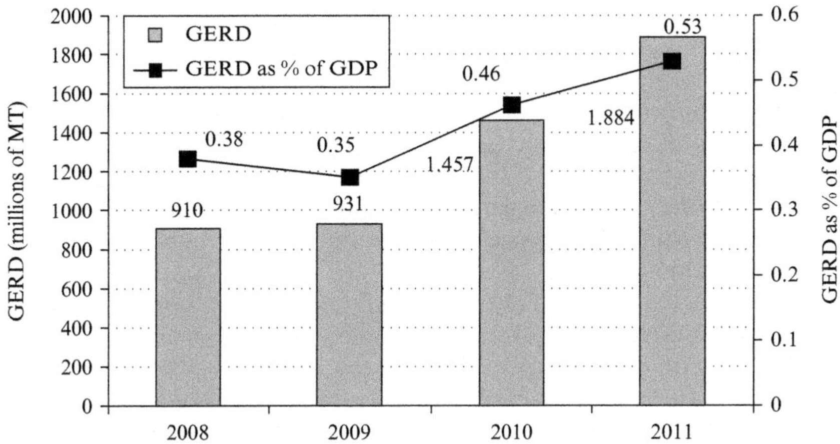

Note: Estimated value, 2011.

Source: MCT national survey, 2012.

Figure 7.4 Gross expenditure on R&D as a percentage of GDP in period 2008–11

as diverse as institutions and individual researchers. Other funds available to researchers include the UEM research fund and the Quality and Innovation Fund of the Ministry of Education and Culture.

FNI has called for tenders to select projects on a competitive basis. In this context, FNI funded 209 projects in total in the period 2010–13 of which 65 per cent are research projects, 56 per cent innovation and transfer of technology projects and 8 per cent aggregate innovation projects, several containing both research and innovation elements. The available financial resources do not respond to the high number of requests for funding from institutions and researchers. As example of this situation is that 617 projects were submitted to NRF in the period 2010–13 but only 171 projects were approved and funded, representing 28 per cent of the total number of project proposals.

Elming and Abrahamsson (2010) have made an in-depth evaluation of FNI as a tool for research capacity building, and provided a set of important recommendations for its improvement. Some of the key findings of this assessment show that the Fund is an important tool to develop science and constitutes an excellent channel for international donors. The overall assessment is that the Fund and its functions respond to what can be expected from a rather new organization both in terms of research administration and management. It is also found that the Fund has high credibility among researchers.

There are some limiting factors that are found in the operation of FNI such as the rather low level of competence of local researchers by international standards, and this has a direct impact on the access to scientifically qualified external reviewers. Further, as most applications are written in Portuguese, the international reviewers' pool is even more limited. Not only is the quality of proposals still low but also the number of female applicants is low as well.

In addition to the State budget and FNI, research institutions can obtain financial resources directly from international partners and open several credit lines with different donors and research agencies; examples include the Open Research Fund from UEM, Institutional Development Fund from the Ministry of Education, Fund for Agricultural Development and the Housing Development Fund.

7.3.6 Productive and Demand Levels

These levels are composed of a large set of public and private actors in the innovation system. Evidence shows, however, that there are poor connections or linkages between companies, companies with research institutions, universities and public administration.

Micro and SMEs make up the majority of the Mozambican business fabric. Official statistics indicate that the micro and SMEs constitute more than 90 per cent of the total businesses and the remaining percentage is composed of large companies and megaprojects.

SMEs are the largest generators of employment and income and the main drivers of economic development. However, the growth of micro and SMEs is limited by obstacles such as lack of a facilitating business environment (heavy taxation, high licensing costs, asymmetric information and inefficient public administration) and difficult access to credit and complexity in the application process. The greater part of micro and SMEs is dedicated mainly to trade and is not in a position to invest in developing products and new processes to be placed on the market.

FDI, on the other hand, has greatly increased since the 2000s after two megaprojects were initiated, the Mozal project for aluminium, and the Sasol project in the gas sector. Following the Organisation for Economic Co-operation and Development (OECD) review (OECD, 2013), it is clear that these two projects created a landmark change in the FDI trend and catalysed subsequent FDI inflows. By 2011, 83.3 per cent of the total FDI was in the extractive industries. Brazil with 43 per cent, Mauritius with 22 per cent and South Africa with 6 per cent of the total have been the major foreign investors. Also, as analysed by the OECD study, FDI in agriculture has been rising, especially in vegetable oils and bio-fuel processing (following the Strategic Biofuel Policy in its 2009–15 pilot phase).

FDI investments have not only the potential to contribute to economic growth but also create local capacities through technology transfer; however, as the case in many developing countries, this potential contribution in Mozambique seems to be overlooked in favour of short-term political benefits that include, for example, tax exemptions. Such context does not favour better linkages of the investors with local higher education institutions and subcontracting local firms as suppliers.

7.3.7 Research and Innovation Input-Output Main Indicators

A research and innovation survey was carried out in 2012 and collected information up to 2010. The main results from the survey show:

a. Investment in R&D in 2010 amounted to 1,456.99 million Meticais in 2010 (see footnote 4), of which the public sector was responsible for 54.9 per cent, higher education for 36 per cent, enterprises for 8.7 per cent and non-governmental organizations (NGOs) for 0.4 per cent.
b. The public sector component was 66.9 per cent from donors, 22.3 per cent from government budget, 10.5 per cent from other sources and 0.3 per cent credit.
c. The majority of researchers has a BSc degree and is concentrated in the areas of social sciences and the humanities (Table 7.3).

The scientific research outputs as measured by bibliometric indicators are shown in Figure 7.5. It can be seen that there is a noticeable increase in the total number as well as the sector number of publication between

Table 7.3 Researchers by scientific area in 2010

Scientific area	PhD	MSc	BSc	Higher high school[a]	Total
Natural sciences	47	97	130	32	306
Agricultural sciences	51	122	162	7	342
Medical sciences	36	60	113	0	209
Engineering and technology	31	36	64	19	150
Social sciences and humanities	116	215	236	14	581
Total	281	530	705	72	1588

Note: a A special category for students in technical schools.

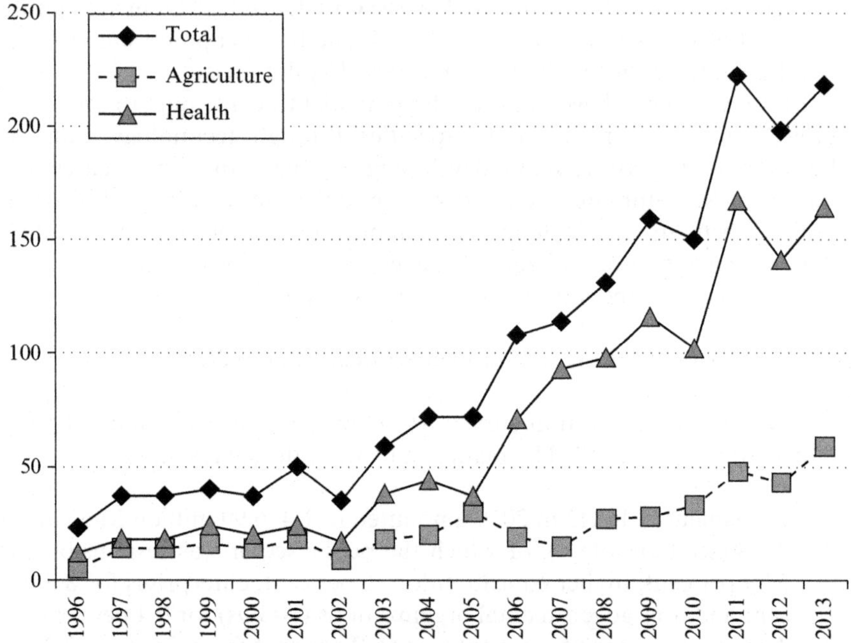

Note: [a] Agriculture includes biological sciences; health includes immunology, microbiology, medicine and neurosciences. The number of papers per sector (for all sectors) when summed up does not give the total number, as some sector papers are counted more than once in the individual search.

Source: SCIMago Lab (2015).

Figure 7.5 Research publications: total in agriculture and health[a]

1996 and 2013. It is possible to attribute this increase to an improvement in the policy context since the 2003 policy was adopted as well as the establishment of MOSTIS in 2008. In spite of the observed improvement, the impact factor is still low when compared to other developing countries (Table 7.4).

Regarding relevance, which is the central issue in 'Closing the Loop', the close policy connection between research priorities and national economic or poverty reduction goals creates expectations that are too high on the social and scientific impact of most projects. The evaluation of Elming and Abrahamsson (2010) points to the fact that even though it is usually fine to expect high impact outputs, there is a strong need for fundamental research that is the basis for creating and adopting new knowledge, methods and techniques.

Table 7.4 Main bibliometric indicators for Mozambique

	1996–2013
H-Index	60
Documents	1762
Citable documents	1678
Citations	21,678
Self-citations	2258
Citations per document	12.30

Source: SCIMago Lab (2015).

Although fundamental research makes up part of MOSTIS, it is not reflected in the guidelines of the Fund. 'The priority on quick results also hampers the development of research and of the researcher, since the results are most often not scientifically good enough to be published in international journals' (Elming and Abrahamsson, 2010).

7.4 THE AGRICULTURE SECTOR INNOVATION SYSTEM

7.4.1 Overview of the Agriculture Sector in Mozambique

Mozambique is a typically agrarian country and development policies and strategies have accordingly awarded high priority to agriculture. About 25 per cent of the country's GDP comes from agriculture (crops, forestry, livestock and fisheries) that represents the main source of income for around 80 per cent of the population. The share of agriculture in total GDP has been fairly stable in spite of the surge of new resource-based industries from around 2000.

As poverty concentrates in the rural areas and an improvement in the standard of living depends on agriculture, the transformation of agriculture is unlikely without an effective and efficient agricultural research and innovation system, and particularly by searching for local technological solutions.

Trends in agriculture development of relevance to this study may be summed up as follows (Benson et al., 2015):

a. Nearly 70 per cent of the total population of the country lives in the rural areas and has agriculture as its means of income. The contribution of agriculture to GNP was relatively stable between 2001 and 2010, having varied between at 24.2 per cent and 25.6 per cent yearly.

b. Vegetable production is 78 per cent of the agricultural GNP. The subsectors of fisheries and forestry represent 7 per cent and 9 per cent, respectively. Other crops such as cotton, tobacco and cashews are also of importance but represent a smaller fraction of the contribution to GNP.
c. The larger part of agriculture is carried out by small farmers or family agriculture. The latter is the main source of employment. Between 5 per cent and 10 per cent of small farmers use improved seeds and 10 per cent animal traction. Only 5 per cent of farmers use inorganic fertilizers and the country has only used 2 per cent of its irrigation potential.
d. There are agricultural activities of large dimensions at the commercial level. Horticulture and rice in one region, tea in another and yet sugar in a third region are typical. The latter crop has attracted foreign investment and some large farms are owned by foreigners. A particular case is that of the production of bio-ethanol and bio-diesel.

Agricultural policy has followed the Plan for the Strategic Development of the Agricultural Sector (PEDSA) (2010–20), which followed two other earlier programmes, PROAGRI I and II. An important contributing strategy was the Strategic Plan for the Development of the Mozambique Agricultural Sector (PNISA), with its origin in the Comprehensive Programme of African Agricultural Development (CAADP) initiated in 2011. The evaluation of Benson et al. (2015) of both PNISA and PEDSA consider that the former provides an adequate base for reaching the goals of the latter.

Mozambique's National Programme for Agricultural Development (PROAGRI) was developed in the mid to late 1990s as an attempt to address existing problems in the sector. According to PROAGRI's Master Document, the programme was expected to (1) develop a common vision for national agrarian development with better harmonized interventions by the State and other partners operating in the sector; (2) be more effective and efficient in the use of public resources (domestic or external) in the sector; (3) reduce the negative impact of project orientation; and (4) reduce dependence on external technical assistance (MAP, 1998).

One result of the programme was a significant increase in aid to the agriculture sector (Pauw et al., 2012). In the case of PROAGRI I, between 1999 and 2005 donors supplied 88 per cent (USD 162 million) and the government 12 per cent (USD 22 million) of the resources. Of the committed sum, USD 152 million was actually spent on specific programme components. A particular emphasis was placed on institutional development and the improvement of institutional processes, such as the establishment of programme management instruments and coordination. The focus on institutional development can be clearly seen by the fact that actual expenditure compared to planned expenditure more than doubled to reach 43 per cent of the total PROAGRI budget (Ministry of Agriculture, 2007, p. vii).

Investments were also allocated to other areas such as livestock production, extension, agricultural production, forestry and wildlife, land management, research and irrigation. Many institutions and aid agencies were among the major contributors to PROAGRI I such as USAID, European Commission, Danida, SIDA/SAREC, UK Department for International Development (DFID), World Bank, AUSAID, Internation Fund for Agricultural Development, United Nations Development Programme (UNDP), Food and Agriculture Organization (FAO), FINNIDA and CIDA (World Bank, 2006).

PROAGRI I was an ambitious programme with high goals. In 2006, an independent consultant was commissioned by the Ministry of Agriculture to evaluate the results. The evaluation showed that PROAGRI was a successful programme, particularly in the area of capacity development at the ministerial level. By establishing instruments of programme management and coordination, it succeeded in building a shared vision of agricultural sector development. Furthermore, new and effective mechanisms for co-ordination between the ministry and the donor community replaced the fragmented donor-driven activities in the sector (Ministry of Agriculture, 2007).

The subsequent PEDSA programme offers a vision by which the subsistence agriculture of today is transformed into a competitive and sustainable sector and establishes a growth objective for the sector of 7 per cent per annum. The calculation of Mogues and Benin (2015) shows that between 2000 and 2011 this growth exceeded the former goal as it reached 8.4 per cent yearly, mainly due to the expansion of agricultural areas.

The above figure is challenged by Pauw et al. (2012) who estimate the growth of the sector at only 3.4 per cent. Further, it is indicated that the public expenditures in agriculture are not applied in sectors that can improve agricultural areas and labour productivity.

Expenditures for R&D and support to farmers and other agriculture services have amounted to no more than 25 per cent of the total expenditures in the sector. Further, the relation between public expenditures and agricultural R&D and the agricultural GNP are low, of the order of 0.25 per cent compared with 0.61 per cent, the average for Africa (Stads and Beintema, 2012; World Bank, 2011).

In the analysis of Mogues and Benin (2015) it is considered that R&D suffers a large temporal gap between the time a specific R&D effort is initiated and the time its outputs are put in the hands of farmers. This problem constitutes a disincentive to the link that should exist to an adequate and stable provision of funds for R&D within a systemic view.

7.4.2 The Agriculture Sector Innovation System

The structural adjustment policies in the 1990s strongly impacted the agricultural sector. Privatization, reduced State influence, falling investments and a high degree of donor dependence necessitated new strategies for the development of the sector (Cabral, 2009). In late 2004 the government of Mozambique established the Institute of Agricultural Research of Mozambique (IIAM). IIAM combined the staff and mandates of the then existing five separate organizations: National Institute of Agronomic Research (INIA), Institute of Animal Production (IPA), National Institute of Veterinary Research (INIVE), Centre of Forestry Research (CEF) and Centre for Agricultural Training (CFA) (IIAM, 2006). The merger of the six institutions was hoped to lead to economies of scale in R&D as well as a more efficient use of resources by better utilization of the main agricultural experimental stations in targeted geographical zones (Walker et al., 2006).

Three higher education centres, UEM, with two faculties, Faculty of Agronomy and Forestry Engineering and Faculty of Veterinary Science, and Mussa Bin Bique University conduct research. More recently other universities have joined the R&D effort, the Higher School of Marine and Coastal Sciences (attached to UEM), Faculty of Agriculture of the Catholic University and Higher School of Rural Development affiliated with UEM.

Other non-profit or private companies collaborate with IIAM and Mondlane University but actually have minimal involvement in agricultural R&D (Bailey et al., 2011; Flaherty et al., 2010). There is also an extensive collaborative process taken jointly with centres of the Consultative Group on International Agricultural Research and other research organizations in developed countries.

Financial and human resources

Table 7.5 provides an overview of the agricultural R&D spending. The expenditures in R&D as a share of GDP have varied from 0.55 per cent in 2004 to 0.35 per cent in 2011 (Nhamusso, 2013).

Flaherty et al. (2010) observed that agricultural R&D is highly dependent on donor funding, which generally fluctuates from year to year, to which should be added that in 2008 government support for research decreased considerably and coupled with budget restrictions and late and irregular

Table 7.5 R&D spending (million 2005 purchasing power parity (PPP) dollars)

Agency	2004	2005	2006	2007	2008	2009	2010	2011
IIAM	9.8	14.0	12.0	9.4	10.5	14.0	16.0	13.8
IIP	3.6	3.6	3.8	5.4	3.7	3.7	2.5	2.7
Higher education	4.0	4.2	4.1	3.8	3.8	4.1	4.2	4.1

Note: Figures estimated by the authors from Nhamusso (2013).

Table 7.6 IIAM and IIP funding share (percentage)

Agency	2004	2005	2006	2007	2008	2009	2010	2011
Government	38	32	35	42	44	38	44	60
Donor	62	68	65	58	56	62	56	40

Source: Nhamusso (2013).

disbursement of funds, the performance of agricultural R&D has been negatively affected. Table 7.6 compares the contribution of funding from government and donors for both IIAM and IIP between 2004 and 2011.

One source of funding between 2007 and 2010 has been the Competitive Grants Programme for Agrarian Research (COMPETE), run with the support of USAID and IIAM. It operated with a budget of USD 2.4 million and most of the funds were allocated under the form of direct agricultural research grants.

The number of full-time researchers (FTE) for 2011 is shown in Table 7.7.

Table 7.8 shows the qualifications of researchers in the agricultural sector in 2011. At IIAM there were 18 PhDs, the largest number in 2011; ten researchers belonged to five universities. The Swedish International Development Cooperation Agency is for UEM the largest donor in terms of training, contributing around USD 6 million per year.

Despite the overall capacity growth recorded, average qualification levels did not improve as much. This is mainly due to prolonged recruitment restrictions, inadequate training opportunities, non-replacement of retired and/or departing senior scientists to private companies and finally, the exceeding the capacity for appropriate mentorship by senior researchers.

The effectiveness of agricultural R&D is influenced by how financial

Table 7.7 Number of FTE researchers in the main research institutions

Agency	FTE
IIAM	210
IIP	38
IS Gaza	12.6
IS Manica	15.5
UEM	
Faculty of Agronomy	20.7
Faculty of Veterinary	13.5
Higher School of Marine Sciences	3.3
Total	313.6

Source: Nhamusso (2013).

Table 7.8 Researchers by degree (public sector)

Degree	2004	2005	2006	2007	2008	2009	2010	2011
PhD	20	20	20	25	25	20	25	30
MSc	60	60	62	67	70	75	78	100
BSc	85	105	170	180	180	185	185	220

Note: Estimated by authors from Nhamusso (2013).

resources are allocated across salaries, operating costs and capital investments. Detailed data on cost categories of government agencies were collected as part of the study of Flaherty et al. (2010). In 2008, about half of total agricultural R&D expenditure at IIAM was spent on salaries, while operating and capital costs accounted for 47 per cent. Compared to previous years, this allocation of resources reflects the higher wage bill incurred by an increasing number of researchers. By 2011 such structure had not changed much; the study of Nhamusso (2013) shows that 54 per cent was spent on salaries, 33 per cent on operating and programme costs and 13 per cent on capital investments.

Increasing the numbers of researchers does not necessarily lead to increased efficiency without improvements in institutional processes. In Mozambique, the higher education sector is plagued by bureaucratic procedures and an overwhelming teaching obligation, affecting the efficiency and quality of research in Mozambique (Hydén, 2006). For instance, at UEM, which ten years ago had only a few thousand students, this number has now skyrocketed tenfold, particularly at graduate level. Consultancy for international organizations and donor agencies is another source of distraction, especially for senior researchers who can supplement their comparatively meagre local salaries with higher consultancy fees (Hydén, 2006).

Moreover, researchers in agricultural sciences in Mozambique have little impetus to cultivate linkages outside their traditional research environment. Professionals tend to prefer interacting with colleagues and attend academic meetings among themselves, rather than develop linkages with non-traditional actors, for example, from the agro-business community, farmers' organizations or NGOs. Examples of such local or international innovation networks, where innovation is more likely to occur, are few in agriculture in Mozambique (Davis et al., 2007).

Flaherty et al. (2010) note that Mozambique's agricultural R&D capacity levels on the whole increased between 2004 and 2008 with the flow of increasing funding and new researchers. Compared to other African countries, researchers in Mozambique are younger and, as a group, have less formal training as researchers. In 2011, researchers holding a BSc certificate as their highest degree accounted for half the total researchers in the country and over 60 percentage of researchers at IIAM and IIP (Table 7.8).

Main research lines

Between 2008 and 2011, the predominant focus of agricultural research in Mozambique was crops and fisheries. The former represented 35 per cent (2004) and 40 per cent (2011) of the FTE researcher's efforts, while the latter represented 20 per cent and 10 per cent, respectively. Livestock research increased from 15 per cent to 32 per cent in the same period; forestry decreased from 10 per cent to 5 per cent; natural resources had a small share of around 5 per cent (crops) and 3 per cent (fisheries) in the period. Other research represented 15 per cent to 10 per cent over the same period (Nhamusso, 2013; USAID, 2008; World Bank, 2012).

7.5 THE HEALTH SECTOR INNOVATION SYSTEM

7.5.1 Overview of the Health Sector in Mozambique

The health sector of Mozambique can be broadly characterized as follows:[7]

a. The Poverty Reduction Strategies have contributed substantially to reduce the number of poor and there is promising progress in some health-related MDGs; however, 54 per cent of the population still lives below the poverty line and access to basic social services remains low, especially for women and children and in rural areas.

b. In spite of progress, several health outcomes are still unsatisfactory, principally malaria, HIV and tuberculosis.
c. There is still inadequate financing, shortage of health professionals and essential medicines. The individual and public health consequences of chronic non communicable diseases, neglected tropical diseases, road traffic injuries and the hidden tragedy of domestic violence need to be addressed better through the public health system.
d. Emergency preparedness and response, including stronger surveillance systems and implementation of the International Health Regulations, require strengthening.

The health policy framework for Mozambique is articulated through the Five-Year Government Programme (2010–14), the Action Plan for the Reduction of Poverty (PARP, 2011–14) and the National Economic and Social Plan (2014). A new Health Sector Strategic Plan 2014–19 was approved following a comprehensive review of the previous 2007–12 Strategic Plan. The Sector Strategic Plan comprises seven strategic objectives based on principles of primary health care, equity and better quality of services.

The health system is composed of a public, a private for-profit and a non-profit private sector; the public sector is the main provider with a network covering about 60 per cent of the population. Strengthening this system is one of the above-mentioned strategic objectives.

Mozambique has more than 25 bilateral and multilateral cooperation arrangements to support the health sector, of which the United Nations system is strongly present. The country also adopted a 'Sector Wide Approach to the health' (SWAp) in 2000 aiming to improve coordination of external assistance towards better performance of the sector, strengthening government leadership and putting greater emphasis on policy and strategy.

There are several health issues and determinants related to health, such as food security, safe water, sanitation and others that can have a substantial contribution from research, technology and innovation. Such contributions need to be acknowledged in the different plans and strategies in a very explicit way and receive full support to be able to contribute effectively to improving the health system in the country.

7.5.2 Policies and Strategies

The guiding lines for defining policies and strategies in the health sector have been given by the policy adopted in 2003 and MOSTIS, as discussed above. For the specific sector, in 2001 the government of Mozambique approved its Strategic Plan for the Health Sector (PESS) 2001–10, which was almost immediately put into implementation. The Plan contains a specific section on 'research and investigation' and several other sections that call for the intervention of technology developments appropriate to the country.

PESS recognizes that health research 'is in a relatively dormant state with an uncertain agenda, with no clear, concrete guidelines from the Ministry of Health (MISAU) on health research, and no legal framework adopted that defines the sector's mandates and priorities in this area'. It also recognizes that because research activity is given little or no priority, decision making is not research based.

At the time of adoption of PESS human, financial and material resources allocated to this area were severely limited, as well as health system studies. The established policy in PESS states that '[R]esearch should be an important and continual activity that should support the objective evaluation of health policy and program decisions. Collaboration is needed between all the institutions that are involved in clinical, epidemiological, and anthropological and health system research. There should be a balance between different types of research according to the country's main health problems. The National Health Institute should play a leading role in developing health research.'

The Plan established a set of strategies for the development of research. Those that stand out are:

a. The transformation of the National Health Institute (INS) into an organ with greater autonomy and responsibility for directing research in the country, adopting modern management mechanisms.
b. The analysis of the relevance of the research and broadcast the results of the major studies.
c. The creation of a forum within the health system, coordinated by INS to facilitate the implementation of the sector's research policies. The forum acting also as a mechanism for disseminating results and sources of information.
d. The encouragement of the creation and support of the training of groups or NGOs that are interested in carrying out research that is in line with the Ministry of Health (MISAU) priorities.
e. The development of INS in facing its new challenges.

f. The specialization of different research bodies in specific areas of the sector's research agenda, strengthening their comparative advantages.
g. The promotion and development of methodological capacity, mainly for research at Provincial Health Directorate (DPS) level related to their main problems.

Several measures and strategies have been taken after the adoption of PESS that have permitted the strengthening of the health sector. An example of relevance to this study is the launching in 2008 of the Health Workforce Development Plan for 2008–15 as a response to address issues such as lack of training capacity, recruitment bottlenecks, low pay, low motivation, misdistribution, 'brain drain' to the private sector, limited management capacity and a severe absolute and relative deficit of health workers. The plan identifies key strategic targets and indicates scheduled activities to be carried out to achieve them. One of the main areas of focus of the plan is on improving the capacity of the Ministry of Health affiliated training institutions.

In previous years other policies and strategies either directly or indirectly related to the health research system have been adopted, including the National Research Agenda on Human Resource for Health (2012) and the National Health Research Programme (under development in 2014).

Although recognizing the above policy contexts in the health sector, respondents to the 'Closing the Loop' survey (discussed in detail in Section 7.6 below) identified a gap between what is aimed at policies and what is actually being implemented. Respondents noted that although the government funds research through FNI, its impact is still limited and the programme has yet to stimulate researchers to focus on priority areas; most perceptions are that the schemes and incentives for research considerably lack transparency.

There is also the perception that the pipeline through which the process of allocating funds for research is complex and often results at the end in researchers not being able to access funds at all. The bureaucratic burden of support schemes is a significant complaint of researchers. In addition, the Ministry of Science and Technology is considered at a distant level from the contexts where research is to be conducted. It is also not clear whether the health issues benefit from a quota in the scheme of financing or the projects have to concur within a basket of other sectors' priorities. Illustrating respondents' perceptions about research policies in Mozambique, one interviewee noted, for example, that 'there is lack of knowledge [of policies] . . . we feel like there is no communication'.

Respondents recognized that a number of broad policies were developed at the macro level (MCT). However, they also point out that these policies lack health sector adaptation and specificities, or when they do exist have not been shared horizontally amongst research institutions. The exception is the recent attempt to develop a health research agenda with wide stakeholder consultation, which is recognized as a positive initiative by several of the respondents.

7.5.3 Actors in the Sector Innovation System

The health sector innovation system is composed of a number of institutions, led by the Ministry of Health. The most important actor in the system is INS (today fully autonomous), which is the largest doing research (with three research centres), dating back many years and having acquired expertise in laboratory areas, implementation/operational research and epidemiological research on most prevalent infectious diseases, such as resistance to malaria treatment, cholera, HIV and related diseases. INS conducts research in a very wide variety of fields and at the same time promotes health research through its research fund open to calls. It also has its own journal and manages a virtual library.

Another important centre in the sector system of innovation is the health research centre Manica, created in 1996. Manica promotes and conducts biomedical research under a bilateral cooperation programme between the governments of Mozambique and Spain, and with the support of the Hospital Clinic/Universitat de Barcelona (through the Fundació Clínic per a la Recerca Biomèdica).

Other actors of the sector system are the Medical Faculty of Eduardo Mondale University, the Maputo Central Hospital – which has carried out research in some clinical areas, the Higher Institute of Health Sciences, the Pedagogical University and the Regional Centre for Health Development. Other university institutions carrying out health research include: ISCISA, ISCTEM, UCM, UniLúrio, Unizambeze, IS Politécnica,

Many of these institutions carry out or contract their projects, in coordination with the Ministry of Health, with several cooperation agencies, for example, DFID Aid, CDC, Irish Aid, NORAD USAID, Swiss Co-operation, United Nations Population Fund, Save the Children, World Vision and Médecins Sans Frontières. There are other institutions that do not work exclusively with the health sector but which carry out research in certain health areas, such as FAO and the National Statistics Institute.

In addition, there also exist a large number of NGOs (such as JHPIEGO, CHAI, HAI, Mchip, ICAP, AGPAF, EGPAF, FGH, ICRH, IESE, CIP), councils (such as AIDS, Social Determinants of Health and SETSAN), foundations (such as the Maniça Foundation and EQUINET) and committees (such as the National Committee for Bioethics in Health and institutional research ethics committees).

INS is conducting research as its main mandate from the Ministry of Health. The research conducted by this institution is mostly commissioned by the Ministry. On the one hand, the latter context allows a better translation of findings to policy makers, but at the same time the context does not create an environment in favour of criteria-defined research agendas nor development of adequate mechanisms for funding allocations for research.

In relation to the 'Closing the Loop' approach, there is an envisioned role for INS that may be summarized in five aspects:

a. Act as a research sector research interface, defining 'priorities' and policies, guides, alignment to agendas.
b. Participate in evaluation and monitoring of quality, practical usefulness and robustness of the health sector research.
c. Conduct research training and *de facto* research focused on major national interest.
d. Integrate actors in health research networks and health system operations.
e. Participate in and promote translation of research findings.

For this role to be fulfilled it is required that:

a. There exists consensus and identifying true synergies.
b. Technical capabilities (clinical trials, implementation pilots, operational research and cost-effectiveness studies) are better developed.
c. Policy dialogues and briefs are produced and disseminated.
d. Capabilities for influencing the use of evidence-based criteria – research – on health policies are developed among researchers and decision makers.
e. Regulatory frameworks are strengthened and timely on health research and knowledge management.
f. There is tracking of health research budgeting.

Respondents to the survey in the health sector noted that research is largely conducted at health sciences universities and the INS. At universities, research is mostly a matter of the academic accomplishment of students as partial criteria to graduate. In such a context, research topics are not necessarily linked to a priority agenda for the university; instead the topics are the priority of departments and supervisors, which to some extent may overlap with national priorities, but invariably is restricted to the health science university branch.

UniLurio University, for example, is focused on nutrition and optometry research only. ISCTEM is mainly focused on oral health. The Faculty of Medicine of UEM is largely focused on epidemiology of diseases. There is also lack of translation of university-led research for policy makers and even for peer-reviewed publication, thus most findings, even relevant ones, remain in the grey literature.

Respondents also noted that NGOs are conducting some crucial implementation and operational research. Nevertheless, their focus is on data collection that is relevant for their supported vertical health programmes. The intent is often to feed the information needed for the health programmes, often vertically delivered, in a context of an unreliable routine health information system. The information generated is largely not fed back to local interested parties. The involvement of local health authorities in NGO-led research is often cosmetic, limited to the bureaucratic matters of the research phases.

Donor-commissioned research (for example, by CDC, UKAID, DFID, USAID) constitutes a considerable part of research conducted in the public sector and within the Ministry of Health, either through INS or NGOs. It is still common that donors' predefined research topics are financially better off, however, there is a sense that nowadays donors are consulting more with the Ministry as a process to define research topics.

Pro-profit NGOs are also emerging as consultancy firms to conduct commissioned programme evaluations and implementation research. The local business sector as such is not involved directly or indirectly in the health sector research. Even so, some pharmaceutical-led evaluations and interests are being commissioned in the country, mainly in the area of infectious diseases diagnosis and treatment.

7.5.4 Research Outputs, Relevance and Impact

Some specific outputs of research carried out locally have been identified that evolve from descriptions to relatively complex processes. Those solutions having high impact have been sought from testing and research including: vaccine trials (malaria – Maniça; HIV – INS; UCM – Beira, Chókwe, Maputo); point of care testing trials; parasites drugs; trials (malaria, schistosomiasis, filariasis); tuberculosis trials (INS – Maputo); increasing systems research (CIOB, FM – UEM).

It is also recognized that there is evolving academic education at a higher level and there exist operational research units at the provincial level.

7.6 SURVEY RESULTS

7.6.1 Introduction

In 2014, researchers and academics active in the areas of health and agriculture in Mozambique were asked to participate in a survey. The survey had as its main objective to elicit the perception of the respondents on the relevance and impacts of research carried out in the national innovation system and, in particular, its policy impact. For this purpose, 169 of a total of 551 researchers active in the fields were selected and asked to participate in the survey;[8] 63 responses were received, 12 from researchers active in agricultural sciences and 51 from health researchers. In the agriculture sector the respondents came from the two main research entities for agriculture, the Agricultural Research Institute (IIAM) and the Eduardo Mondlane University. In the case of the health sector, responses were collected through interviews with key informants in Maputo as well as an online questionnaire reaching into the provinces.

The majority (53 per cent) of the respondents to the survey hold a BSc degree, 32 per cent hold an MSc degree and the remaining 15 per cent hold a PhD degree; the gender distribution of the respondents was almost even with 32 women and 31 male participants.

7.6.2 Quality and Relevance of Research

The opinion expressed by the majority of the surveyed researchers, both male and female and independent of academic degree, on the quality of research in the country is that it is good or average. Over 80 per cent of health researchers and all of the researchers in agriculture hold this view. In the health sector, a minority of 16 per cent disagrees with this assessment and instead characterizes the level of quality as bad (Table 7.9).

Notwithstanding the generally positive evaluation of the quality of research, it is emphasized by the respondents that in order to improve quality there is a set of prerequisites that need to be fulfilled. These include more funds for research, investments in infrastructure and improved human capital for research. Some of the reasons behind the difficulties for improving the quality of research lie in the critical need of higher education to build advanced human capital. Decades of underinvestment together with fluctuations in yearly investment levels in R&D are considered the main reason behind the weak situation of the agriculture sector innovation system.

Also in the health sector, the quality of projects is considered medium and that outputs are not easily transferred to firms or other institutions,

Table 7.9 What is your overall assessment of the quality of research in your country? (percentage)

Research area	Good		Average		Bad	
	Frequency	Per cent	Frequency	Per cent	Frequency	Per cent
Agriculture	2	16.7	10	83.3	0	0
Health	12	24.0	30	60.0	8	16.0

Source: Survey.

thus limiting their use. Projects are felt to be in general inadequately funded and this is one reason for the medium rather than top quality of projects. Further, it is considered that the available funding is difficult to access and that it is necessary to minimize the dependence on external funds.

In order to improve the quality of research in the innovation system, there is also a stated need for an increase in investment in equipment and infrastructures and a strategy that would include an increase of salaries, which today are inadequate and uncompetitive in the opinion of the researchers.

The low incentive for research is reflected in the fact that there are few full-time researchers. Time dedicated to research varies greatly, but at least 50 per cent of scientists in agriculture invest less than 50 per cent of their time in research. This is one indication that the research and the innovation system are not performing properly.

In the health sector, the opinions on relevance and quality of research projects ranged from 'pertinent, rigorous and of great impact' to 'exclusively . . . by Mozambicans . . . methodologically weak, especially statistics'. From the perspective of outputs in the form of scientific papers, the majority of good scientific results is published in English journals. There is also a preoccupation that research is duplicated. Although this is a debatable issue, it is believed that it is a limitation for the use of results, as they may differ from one project to another.

Respondents to the survey emphasized that despite questions of local technical capacity to conduct research, relevant research is conducted in Mozambique. In addition, HIV, malaria and chronic diseases such as under-nutrition start to emerge as important research topics. Interestingly, a prevalent notion was that tuberculosis is a topic lagging behind research, when compared to non-communicable chronic diseases, despite the fact that it is closely associated with highly researched HIV.

7.6.3 Funding Policies of Research

Researchers feel that the funding systems are not working properly. More than 80 per cent of the researchers do not feel confident that the actual system performs as it should. Funding is identified as a key issue for 'Closing the Loop' and, as already mentioned above, pivotal for the quality of research undertaken in the innovation system. From the opinion of the experts consulted in this study, some of the suggested measures to improve the funding situation can be summarized as follows:

a. Funding for research should be increased and allocated in order to encourage and stimulate researchers in their work; further, funding should be made according to local priorities.
b. Encouragement also comes from improvements in infrastructure and work conditions; in particular, there should be better incentives for researchers, particularly when they show good performance. It is also necessary to provide incentives for junior researchers.
c. Transparency in the criteria for research funding with better coordination between research themes and major financers of the agrarian sector. The fact that a research proposal can be dismissed without any feedback to the proponent strongly discourages researchers. It is required that information flows more readily and openly.
d. A strategic plan should be defined for research areas considered crucial for institutional development and according to the government lines of priorities; this plan should consider investment in infrastructure and training human resources. A priority line of research should be on traditional knowledge as part of the scientific heritage built through time in the country.
e. The planning, execution and management of finances in big R&D institutions should be more independent from the respective ministry. Because of a strong dependence, budgets are underestimated and disbursement of funds is often late.
f. The planning, implementation and financial management of large research institutions should be autonomous and not linked to a specific sector ministry to avoid underestimation of research in the face of other urgencies and to allow and promote more interdisciplinary and multidisciplinary research.
g. There should be different levels of funding and criteria, such as the size and importance of the problem/issue to be addressed, research capacity, seniority of the principal investigator and so on. In addition, the postgraduate system in universities should be part of the funding system because research in universities is largely connected to Master's

and PhD degrees. Thus, funding should consider tuition fees, scholarships and costs of the research project.

h. Improve the monitoring system in order to supervise the money application process.

7.6.4 Impact of Research

According to the responses to the survey, there is a lot of relevant research being performed and validated in the country. However, most of the developed technologies or other research results are not used and disseminated to targets groups to their full extent.

Research is an expensive activity. It is thus important to have a well-articulated programme in accordance with government goals and aligned to development policies. The government's ambition is to gradually increase research funding to 1 per cent of GDP but is still falling behind this goal in spite of progress made during the last ten years. As a result, as already stated several times, most of the research funds allocated to R&D systems come from external donors.

This has implications because most investors and donors have their own interests and visions that are not always compatible with those of the government. Non-government funds are typically available internationally through competitive calls. To have access, it is important to obey the criteria, rules and, most important, the research interests and fields determined by the donor, which do not always coincide with the national programmes.

There is disagreement in the opinion of researchers regarding such an alignment between funding agencies and government policy instruments. Around 48 per cent of the researchers believe that this alignment exists while 52 per cent do not see any kind of connection. Researchers in the agriculture sector are slightly more pessimistic than their colleagues in the health sector (Table 7.10). The view in the health sector is that locally funded projects generally are relevant and responding to national priorities, whereas external funds, which are predominant, are not necessarily aligned with such priorities. In particular, it is considered that research commissioned by the Ministry of Health results in more usable outcomes. Moreover, respondents to the survey believe that in order for research to have a better impact, its outputs should be presented in simpler language suitable for end-users.

Many researchers feel that there is an imbalance between applied and basic research, probably a consequence of the fact that the budget allocated to research is very small. It is very difficult to receive funding for basic research from non-governmental sources since most donors only fund applied research. As an overall consideration, researchers in

Table 7.10 Do you think adequate measures are taken by the funding agencies to ensure alignment of research objectives with national developmental goals?

Research area	Yes		No	
	Frequency	Per cent	Frequency	Per cent
Agriculture	5	45.5	6	54.5
Health	22	48.9	23	51.1

Source: Survey.

agriculture feel that there is not much interest in basic research from funders. Basic research has to be covered by the government's limited funds, within a national strategy, enforcement and vision. With increasing funds specifically for the purpose of basic research and with targets and timeframes clearly defined, it is felt that national agricultural research would probably be stronger and more useful for the national industry and economy.

Most of the respondents believe that applied research is very important, and that its results are useful to the sector it is carried out in. The same view extends to the contribution of the research to the economy with a large number of researchers saying that their research is important and has a positive impact on the economy. Over 80 per cent of the respondents in both sectors perceive that their research has achieved impact on local or national levels (Table 7.11).

When asked what kind of impact, 38 per cent reported impact on policy, 38 per cent on social development and the remaining 24 per cent on economic development. In line with the comparatively high level of policy impact, almost 60 per cent of researchers feel that they indeed have a chance to communicate their findings to the policy makers at least sometimes. Adding the 14 per cent of the respondents that feel they always have the opportunity to present research results, almost three out of four researchers have some kind of access to the policy-making process. The remaining 27 per cent think they do not have any access at all to policy makers and consequently have no opportunities to contribute their ideas and thoughts in policy deliberations.

Executed projects have the aim of informing policy and to try to engage policy makers, but these are generally not involved in the implementation of research projects. Despite the reported high degree of impact on policy, research results are not necessarily shared directly with policy makers; opportunities for this to happen are rare. In spite of this limitation, the use

Table 7.11 Do you feel that your research results have had any impact on local or national levels?

Research area	Yes		No	
	Frequency	Per cent	Frequency	Per cent
Agriculture	10	83.3	2	16.7
Health	38	80.9	9	19.1

Source: Survey.

of evidence for policy is increasing as can be verified in government and other public sector documents.

Some researchers are not so optimistic about this situation; they believe that the process is very weak. It is slow, difficult and lengthy. Several existing challenges have been identified regarding the way research is carried out and its outputs informing policy makers, in particular, the need to develop better knowledge management capacity and governance of the sector innovation system. In the health sector, the need for ethics monitoring and the way projects are approved are also recognized as weak.

It is believed that regulatory authority for technologies and trials needs to be reinforced or created as at present there is limited laboratory platforms capacity. Other weaknesses in the health sector system include the low levels of publications (thesis and other research); no patents/inventions for health; financial distortion – research skewed to aid programmes; lack of health system research capabilities; and in general a weak national network for research.

To overcome this situation, it is considered that there is a need for fast updating, especially in the normative guidance that should exist from policy, facilitating research deployment at full scale that will need significant investment on governance and monitoring. Policy and strategy must plan well its implementation; otherwise public health benefits from research may be reduced. Also, policies and guidelines for research prioritization, alignment and implementation need to be developed from a health system perspective.

As seen from the identification of the actors in the sector system of innovation, there are multiple stakeholders, and although it is claimed that the regulatory functions must be improved, there is at present a myriad of regulations.

For research to have better impact it is necessary for a decentralization of power and care into community and informal settings and at the same time appropriate quality assurance and scientific robustness and linkage to health policy needs.

In agriculture, researchers state that laboratories, research network and to a lesser extent research centres are the most important national infrastructure the country has. A well-functioning infrastructure together with appropriate human capacity is a prerequisite for the development of an agrarian research. PROAGRI I andII were put in place precisely to address this issue.

The latter includes infrastructure for information dissemination to end-users. Farmers need information on new technologies available as well as market information to make informed decisions on when, where and how to bring their produce to the market. Most of the information regarding new technologies, and contributing to improved decision making on the productive sectors and the economy, comes from the public sector. Thus, in spite of the growth of the agrarian private sector, most researchers believe that the public sector is the most important and dominating actor in the agricultural sector.

Information from research centres is disseminated primarily through the government's extension networks. In agriculture, the Extension National Network links the farmers to the research systems and vice versa. The Ministry of Commerce (MIC) provides market information designed to meet the needs of traders, whereas the Ministry of Agriculture's (MINAG) Agricultural Markets Information System (SIMA) meets the needs of farmers. With improved information systems, the efficiency and impact of agricultural R&D systems are likely to lead to productivity increases in the agriculture sector in Mozambique, particularly among smallholding farmers.

7.6.5 Strength and Weaknesses of the Innovation System

The researchers interviewed in this survey strongly suggest that the government should be instrumental in linking different sectors of R&D to overcome the 'silo mentality' of line ministries. In order to achieve a properly working R&D system, the researchers suggested that the main role of the government should be to: (a) disseminate research results; (b) support proper application of research results; (c) ensure socio-economic impact of the research in the communities; (d) fund research projects in a competitive way; and (e) stimulate researchers to carry out good research ideas and projects.

The respondents also identified a number of critical issues for the future development of the innovation system. Table 7.12 summarizes the main findings of the survey under a Strength and Weakness analysis.

Table 7.12 Researchers' view on strengths and weaknesses of the innovation system of Mozambique

Weaknesses	Strengths
Institutional: Low transparency in the distribution of the limited resources Lack of funding and/or lack of financial incentives Lack of equipment and material Deficiency in researcher empowerment Low level of autonomy	*Institutional:* A growing, qualified, workforce with recognized researchers per field of knowledge The existence of many research units; even though too small most of them are able to start and guide research projects
Impact: Weak technology dissemination to the farmers Weak network for agrarian inputs Low dissemination rate of innovation Weak connection between research institutes and farmers' enterprises The production value chain is not strong enough to push the agrarian system Nonexistence of a network to identify and stimulate innovation diffusion into the agrarian system Usually the innovation does not address the real necessity of the poor Sometimes innovation shows high costs of implementation	*Impact:* Examples of new areas of production opened Examples of new technologies developed
Policy and coordination: The contribution of a national research network is unknown to the majority of people, usually leading to misunderstanding when it comes to the design of agrarian policies There is no entity that plans, monitors and coordinates the innovation system throughout the country	*Policy and coordination:* There is a political ambition to increase and improve the number of national researchers
Funding: There is no transparency in the process of research proposal evaluation mainly funded by FNI and MCT Sometimes there is a clear dependence on the donors due to lack of national investments Weak financial systems to sustain and stimulate the national innovation system	

Source: Survey.

Reflecting the earlier discussion, most strengths identified by the research community are related to human resources at the institutional level and to what they have achieved in terms of new technologies and products. Conversely, most weaknesses are related to systemic failures in linking different actors of the system as well as in safeguarding that funding is adequate and reaches the deserving projects.

It is also encouraging to note that the research community acknowledges the ambition of policy makers to improve the situation for the research institutes. This indicates that there exists a window of opportunity for a constructive dialogue between researchers and policy makers on the future direction of the innovation system in Mozambique.

7.7 CONCLUSIONS

7.7.1 The National Innovation System: Institutional Issues, Policies and Strategies

The national innovation system in Mozambique is still very weak. The linkages between researchers with other actors of the system are limited. Enterprises of all sizes have limited research capacities and limited capabilities for technological development – which is reflected in their difficulties in attaining quality standards as demanded by the international markets.

Most of the R&D is carried out in public sector institutions. In this sector there are challenges of importance; training of human resources, improvement of the quality of education at all levels, infrastructure and lack of investment in R&D. The government's stated goal is to increase the latter from the present 0.42 per cent of GNP to at least 1 per cent.

Despite encouraging progress, the absence of outcome, result and product indicators and targets in MOSTIS and five-year Government Programme (PQG) constitutes a major obstacle to the implementation of MOSTIS and also for S&T policy. MOSTIS is itself an open space for subjectivity. In the current conditions, the targets were defined according to criteria of different public institutions, and the evaluation and monitoring of MOSTIS are based on the sum of actions carried out by the different public institutions.

The second obstacle has been lack of updating of MOSTIS. Consequently, part of its strategic programme is unsuitable and others have never been implemented since 2008. For example, still to be created are centres of excellence and mechanisms should be found to operationalize the thematic scientific councils of water, energy, health, agriculture and ethno-botany and also the National Council for Science and Technology.

Furthermore, strong institutions endowed with financial and well-prepared human resources are crucial requisites to develop the national innovation system. Evidence shows that most institutions do not have those requisites and are not sufficiently clear on what should be done in the field of innovation. In this context there are many institutions that do not fulfil their role in the system.

Because of lack of qualified human resources (and opportunities for the existing ones), there exist difficulties in understanding and implementing complex programmes, particularly in the fields of biotechnology, ethnobotany, ICT, innovation and others. At the same time, such lack also limits the critical mass of planners and policy makers.

Weak coordination in fund-raising has been one of the obstacles for the development of the innovation system. Budgets in MOSTIS have never been estimated, causing difficulties in the definition of a national strategy for resources mobilization. The current scenario shows that there is no linkage between the different funds and projects implemented by different institutions of the innovation system.

Without good policies it will be difficult to provide guidance to research and innovation, and policies need to be better studied. Further, policy implementation can generally be rated as being poor. Mugwagwa et al. (2015) consider that there is lack of policy coherence, of enforcement and accountability mechanisms, and of financing policy implementation. The authors correctly point out that increased stakeholder involvement and political leadership cannot guarantee policy implementation, unless such limitations are overcome.

Walters-Monteiro (2010, abstract) is very critical of the implementation of the policy efforts in Mozambique, stating that 'despite the emphasis given, in terms of policies, to the importance of science and technology to the development, the practical actions show that the existing habitus in the education system, conditioned by a colonial legacy where the knowledge is transmitted in a limited form and, by the absence of a scientific culture that affects the science and technology system and some segments of Mozambican society, contributes to the maintenance of a vision of small relevance of the scientific and technological activity to the development of the country'.

Such a critical view can actually be extrapolated to situations existing in many developing countries where the rhetoric of the importance of STI is not matched by real commitments to its development.

7.7.2 Sector Innovation System in Agriculture and Health

The critical assessment of the national innovation system given above is also apt for the sector innovation system in agriculture and health. Despite a professed willingness on behalf of the policy makers to put in place policies appropriate for addressing the long-standing challenges of the sector, many problems still persist.

From his study, Nhamusso (2013) concludes that in the case of agricultural research most of the problems are related to finance. There are few national resources for research investment that depends heavily on volatile and short-term donor funding. Further, the study shows that there are many young researchers in need of training.

In the opinion of the research community, the major problem areas are the result of a system failure where the intentions of the often well-formulated and ambitious policies fail to be implemented. One such area is the serious disconnect between the actors in the innovation system where researchers are not sufficiently linked to end-users, such as farmers, companies and the public at large, in order to perform the envisioned services or to develop appropriate technologies. In addition to these weak inter-sectoral linkages, there is a prevalent gap in communications between researchers and policy makers so that research outputs are often not shared with policy makers and opportunities for exchanges of experiences and ideas are rare.

A second problem area pertains to the allocation of resources and mechanisms for funding. Although resources for research in agriculture and health have gradually increased during the last decade, it is clear that research activities are still underfunded in relation to aspirations. In agriculture, the PROAGRI programme has generally been effective in increasing and coordinating donor funding, but the contribution of research to the development of the sector has so far failed to meet expectations. Moreover, mechanisms for distribution of funds are perceived as lacking transparency and with sometimes arbitrary and sometimes politically motivated outcomes. A more transparent and consistent system for funding decisions would improve the prospects for research to realize its potential for supporting development efforts.

The third area, and related to the above, concerns institutional dynamics; capacity levels at research institutes need to be raised and existing competence better utilized. Present incentive structures to some extent work as obstacles to institutional development, as do the low level of autonomy from line ministries. To that can be added a general lack of transparency and feedback mechanism in the process of research proposal evaluation at the ministerial level.

Taken together, the system failure highlights an implementation problem rather than a policy formulation problem. This notion is substantiated in a study of the health systems in Mozambique and Tanzania (Mugwagwa et al., 2015) that finds that although there indeed exists a full set of policies and strategies for guiding research and innovation in both countries, the effective implementation has failed to appear. The authors conclude that there is a need to focus more on the policy implementation process than on developing more policies. Thus, the regulatory framework for STI needs to be strengthened and fine-tuned if Mozambique is to utilize the full potential of research in meeting the needs of the country. A positive factor in such a scenario is the existence of a young, ambitious and steadily expanding research community with growing experiences and skills.

NOTES

* We wish to acknowledge the valuable contribution of João Benedito Carlos Nuvunga and Paula Mussera Viola Salvador for the survey of researchers in the agriculture sector presented in Section 7.6 below.

1. www.worldbank.org/en/country/mozambique/overview (accessed 28 December 2015).
2. OECD Statistical database.
3. The central organ is composed of the national directorates, one of which is the National Directorate of Planning, Statistics and Cooperation (DPEC).
4. The S&T Provincial Delegations and Technology District Services have administrative autonomy (design and implement local activities, plans and budget).
5. To provide an idea of the value of the Meticai in USD terms the following conversion figures may be used: 1 USD = 28,726 Meticais (2011); = 27,928 (2012); = 29,745 (2013); = 31,014 (2014); 38,000 (2015).
6. www.scimagojr.com/countrysearch.php?country=MZ (accessed 27 May 2015).
7. World Health Organization, country strategies, www.who.int/countries/en/ (accessed 15 May 2015).
8. See Chapter 8 for a full discussion of the methodology employed as well as a presentation of comparative results.

REFERENCES

Bailey, T., N. Cloete and P. Pillay (2011), 'Universities and economic development in Africa, case study: Mozambique and Eduardo Mondlane University', Centre for Higher Education Transformation.

Benson, T., T. Mogues and S. Woldeyohannes (2015), 'Sinopse: Determinar os Progressos Efectuados Rumo aos Objetivos Comuns para a Transformacao Agricola em Mozambique', Synthesis of the Position Paper 01370 of the International Food Policy Research Institute, Washington, DC, April.

Brito, L. (2003), 'The Mozambican experience: initiating and sustaining tertiary education reform, a case study for the Regional Training Conference on Improving Tertiary Education in Sub-Sahara Africa', *Things that Work*, Accra, 23–25 September.

Cabral, L. (2009), *Sector Budget Support Practice, Desk Study Agricultural Sector in Mozambique*, Mokoro and London: Overseas Development Institute.
Chilundo, A. (2002), 'The state of higher education in Mozambique', Paper presented at the UNESCO Africa Regional Conference on Progress and Future Directions of Higher Education in Africa, Abuja, Nigeria, 4–7 June.
Davis, K., J. Ekbor, W. Mekasha, C. Ochieng, D. Spielman and E. Zerfu (2007), 'Strengthening agricultural education and training in sub Saharan Africa from an innovation systems perspective. Case studies of Ethiopia and Mozambique', IFPRI, Discussion Paper No. 00736, Washington, DC.
Elming, S.-A. and K. Abrahamsson (2010), 'The National Research Fund (FNI) as a tool for research capacity building in Mozambique: an in-depth assessment of SIDA support to Ministry of Science and Technology MCT within the research cooperation with Mozambique', *SIDA Review*, 15.
Flaherty, K., F. Mazuze and R. Mahanzule (2010), 'Recent developments in agricultural research', ASTI, IFPRI, Washington, DC, October.
Hydén, G. (2006), 'Universities and faculties research funds at Tanzania, Mozambique and Uganda', *SIDA Evaluation*, Stockholm, Sweden.
IIAM (2006), *Priority Setting for Public Sector Agricultural Research in Mozambique with the National Agricultural Survey Data*, Prepared by T. Walker, R. Pitoro, A. Tmo et al., Research report No. 3E, August.
ITAD, FISCUS and MBConsulting (2014), *Independent Evaluation of Budget Support in Mozambique*, Final Report, Vol. I, European Commission.
James, R.C., C. Arndt and K.R. Simler (2005), 'Has economic growth in Mozambique been pro-poor?', International Food Policy Research Institute, Brief Discussion Paper No. 202, Washington, DC.
MAP (1998), 'National Program of Agrarian Development – PROAGRI, 1998 to 2003', Volume II, Master Document, Ministry of Agriculture and Fisheries, February.
Ministry of Agriculture (2007), *Evaluation of the First Phase of the National Agriculture Development Program PROAGRI (1999–2005)*, Vol. 1, Main report.
Mogues, T.T. and S. Benin (with S. Woldeyohannes) (2015), 'Despesas Públicas na Agricultura em Mozambique: Quais os Investimentos Neccessario a Evolucao Tecnologica e o Que Norteia as Decisoes de Investimento?', International Food Policy Research Institute, Mozambique Strategy Support Programme, Note 7, April.
MOSTIS (2006), Mozambique Science, Technology and Innovation Strategy: Time Horizon 10 years, Approved by the Council of Ministers in the 15th Regular Session, 27 June.
Mouton, C. (2007), *A Baseline Study of Science and Technology and Higher Education in the SADC Region*, ed. C. Mouton and J. Mouton, Studies Series, University of the Witwatersrand: Southern African Regional Universities Association (SARAU).
MST (2014), Mozambique Innovation System (MIS), Ministry of Science and Technology, Draft, Maputo, March
Mugwagwa, J., D. Edwards and S. de Haan (2015), 'Assessing the implementation and influence of policies that support research and innovation systems for health: the cases of Mozambique, Senegal, and Tanzania', *Health Research Policy and Systems,* **13**, 21, doi:10.1186/s12961-015-0010-2, available at http://www.health-policy-systems.com/content/13/1/21 (accessed 15 December 2015).
Nhamusso, A. (2013), 'Transformation of agri-food systems and commercialization

of smallholder agriculture in Mozambique', Paper presented at the Evidence, Challenges and Implications; Agriculture R&D Trends and Challenges in Mozambique Workshop, 9–10 December, available at http://wwwasti.cgiar.org (accessed 15 December 2015).
OECD (2013), *OECD Investment Policy Reviews: Mozambique*, Paris: OECD, October.
Pauw, K., J. Thurlow, R. Uaiene and J. Muzonda (2012), *Agricultural Growth and Poverty in Mozambique*, Technical analysis in support of the Comprehensive Africa Agriculture Development Programme (CAADP).
Pivodic, F. and D.S. Nhachengo (2014), 'Is the concept of absorption capacity misleading us? The case of the Mozambican health sector', Paper presented at the Third Global Symposium on Health Systems Research: Science and Practice of People-centered Health Systems, Cape Town, 30 September–3 October.
Stads, G.-J. and N. Beintema (2012), *Agricultural R&D in Africa: Investment, Human Capacity and Policy Constraint. Improving Agricultural Knowledge and Innovation System*, OECD Conference Proceedings, Paris: OECD Publishing.
Torvinen, L. (2013), 'Assessing governance assessments; the case of Mozambique', PhD Dissertation, Maastricht University, 21 November.
UNDP (2014), *Human Development Report 2014*, United Nations Development Programme, New York.
USAID (2008), *Private Investment in the Agricultural Sector in Mozambique*, Washington, DC.
Walker, T., R. Pitoro, A. Tomo et al. (2006), *Priority Setting for Public-sector Agricultural Research in Mozambique with the National Agricultural Survey Data*, Research Report No. 3E, Institute of Agricultural Research of Mozambique, Research Report Series, August.
Walters-Monteiro, M.H. (2010), 'Ciencia e Tecnologia: Producao de Conhecimento em Mocambique entre 2003 a 2008', Master's Thesis, Universidade Federal do Rio Grande do Sul, Porto Alegre, Brazil.
World Bank (2006), *Mozambique Agricultural Development Strategy Stimulating Smallholder Agricultural Growth*, Report No. 32416-MZ, AFTS1, Agriculture, Environment, and Social Development Unit, Country Department 2, Africa Region.
World Bank (2011), *Analysis of Public Expenditure in Agriculture*, Vol. 1, Core Analysis, Report No. 59918-MZ, Washington, DC.
World Bank (2012), *Agricultural Indicators: Mozambique*, Report No. 68241, Washington, DC.

8. Making research matter: a synthesis of survey findings

Bo Göransson

8.1. INTRODUCTION

The four case countries examined in this book differ substantially in terms of size, level of economic development, industry structure and market orientation, but they all share the trait of assigning high importance to the efficient utilization of science, technology and innovation (STI) for development. The preceding chapters have illustrated the diversity and heterogeneity of these attempts, but also that the *raison d'être* for national policies in this area is basically the same; to better align the research that is actually carried out with the developmental needs of the country. In this chapter, we take a closer look at how researchers perceive not only the relevance and quality of their research but also the ability of the national innovation system to put research results into productive use.

8.2 METHODOLOGICAL CONSIDERATIONS

The empirical data have been collected in a project that started in 2010.[1] After extensive discussions with the team leaders of the four case countries, a pilot questionnaire was developed, tested and subsequently revised and translated to the languages of the case countries together with an open-ended discussion manual for semi-structured interviews of researchers, policy makers and private sector decision makers.

The purpose of the questionnaire was to map out the state of the art of the research and innovation system and governance structure in each of the countries, and to elicit the opinions of researchers and policy makers on a number of issues related to the relationship between the research community and the policy-making sphere. How much impact or influence do researchers feel they have on policy making? How free are they in formulating and proposing research projects? And how inclined are they to follow the intentions and objectives of the developmental plans laid out by the policy makers? Likewise, how do the policy makers feel about the relevance of the research result being presented to them? Are they aware of research projects being conducted in their area of responsibility?

In addition to the questionnaire, information was gathered at national workshops bringing together main actors of the national systems of innovation – governments, universities, entrepreneurs and civil society – in the participating countries in a constructive dialogue to further communication between the groups. One national workshop was conducted in each of the case countries.

The survey was conducted by teams of researchers in the case countries. A survey aiming at full coverage of all research areas would have required a larger sample than was possible for the project. For this reason, it was decided to concentrate the analysis on two important areas for development in the case countries: health and agriculture. The target groups for the survey were researchers engaged in scientific research on health and agriculture either at universities or specialized research institutes. For Bolivia, Tanzania and Mozambique this means a headcount of between 209 to 620 researchers (Table 8.1). In each country, a sample of researchers was selected from the total population of researchers. The sample differed in size from 250 in Bolivia to 60 in Tanzania. In total, 241 researchers responded to the survey. The response rates ranged from 30 per cent in Bolivia to 77 per cent in Tanzania where the survey was conducted through face-to-face interviews with researchers at selected organizations. Although the response rate was low also for Mozambique (37 per cent), the

Table 8.1 Sample characteristics of the four surveys

Country	Total target population (HC)	Sample (no.)	Responses (no.)	Response rate (%)
Bolivia	2153	250	75	30
Agricultural sciences	*327*			
Medical sciences	*340*			
Vietnam	105,230	100	57	57
Agriccultural sciences	*n.a.*			
Medical sciences	*n.a.*			
Tanzania	3102	60	46	77
Agricultural sciences	*527*		*16*	
Medical sciences	*620*		*30*	
Mozambique	1588	169	63	37
Agricultural sciences	*342*	*24*	*12*	*50*
Medical sciences	*209*	*145*	*51*	*35*

Source: UIS (2015), Ministry of Science and Technology and national surveys.

survey still managed to solicit the opinion of over one in ten researchers active in agricultural and medical sciences in that country.

The relatively low number of researchers participating in the surveys limits the statistical value of the results. This is particularly so in the case of Vietnam, which includes over 100,000 individuals in its comparatively generous definition of what constitutes a researcher. Yet, we feel that the responses from the researchers as well as from interviewed policy makers and managers provide a valuable and accurate insight into the relationship between researchers and policy makers in the case countries, particularly in the areas of health and agriculture. There is also always a danger in generalizing too much over different research areas as they may face different challenges and realities. But, as the presentations of the empirical material will show, it is clear that many of the experiences that the researchers have had in the two research areas are representative also for the research community at large in the four countries.

8.3 SAMPLE CHARACTERISTICS

Of the researchers participating in the survey, women accounted for a third of the total (Figure 8.1). This fits well with the average number of women researchers in developing countries in general. However, we observe an underrepresentation of women in the samples for two of the countries. As presented in Table 3.1 in Chapter 3, Bolivia's workforce of researchers consists of 67.5 per cent women whereas the sample only included 28 per cent. Contributing to this discrepancy is the fact that many female researchers work part time; the number of full-time researchers is smaller for women than for males. In addition, there are much fewer women researcher in the agricultural sector when compared to the health sector. In Tanzania the corresponding figures were 25.4 per cent versus 15.2 per cent. In Vietnam there is a fairly good fit with 41.7 per cent versus 36.8 while there is an overrepresentation in the sample in Mozambique (32.2 per cent versus 50.8 per cent).

The sample is dominated by the age group 31–44, which accounts for over 50 per cent of the sample, followed by researchers with an age above 45 years, accounting for 35 per cent (Figure 8.2). Younger researchers are included in all the country samples but, with the exception of Vietnam where they account for almost a third of the respondents, to a minor extent. The lower response rate from that group could partly be explained by a general reluctance of junior researchers to respond to questions they perceive as 'above their paygrade' and experience.

The educational level of the respondents in the survey is fairly equally

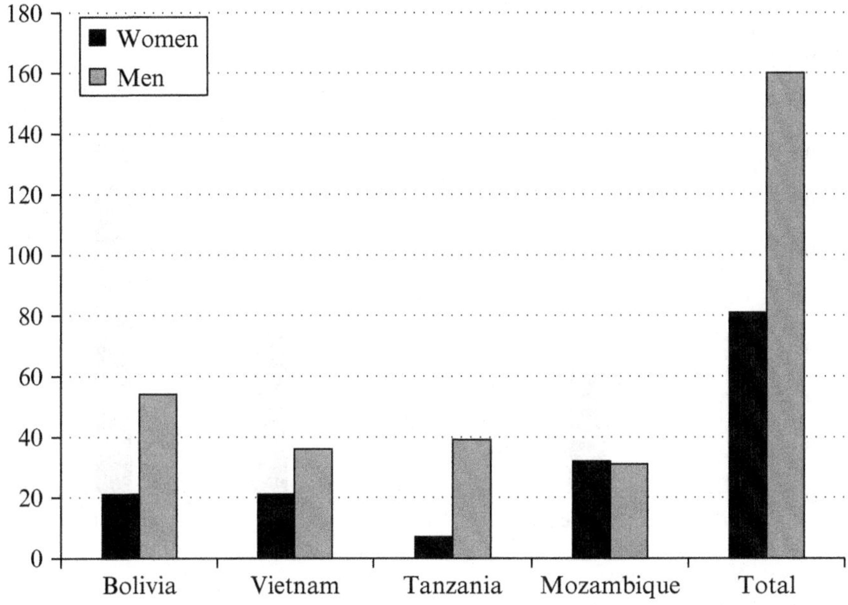

Source: National surveys.

Figure 8.1 Gender sample distribution by country (number)

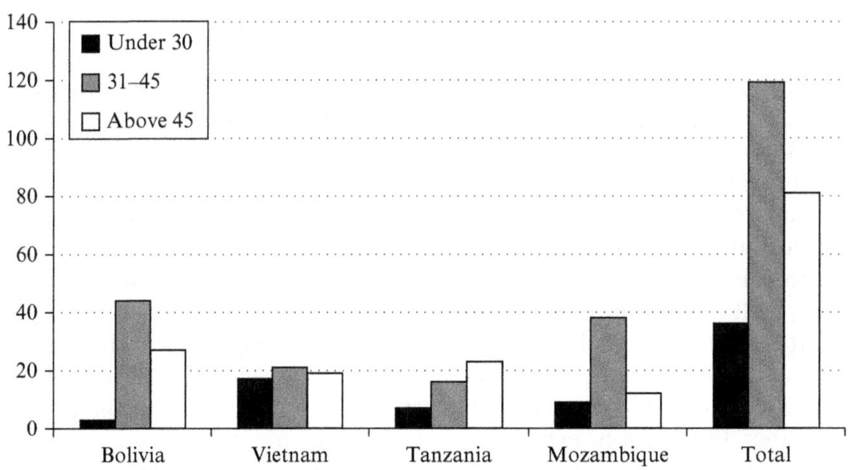

Source: National surveys.

Figure 8.2 Age sample distribution by country (number)

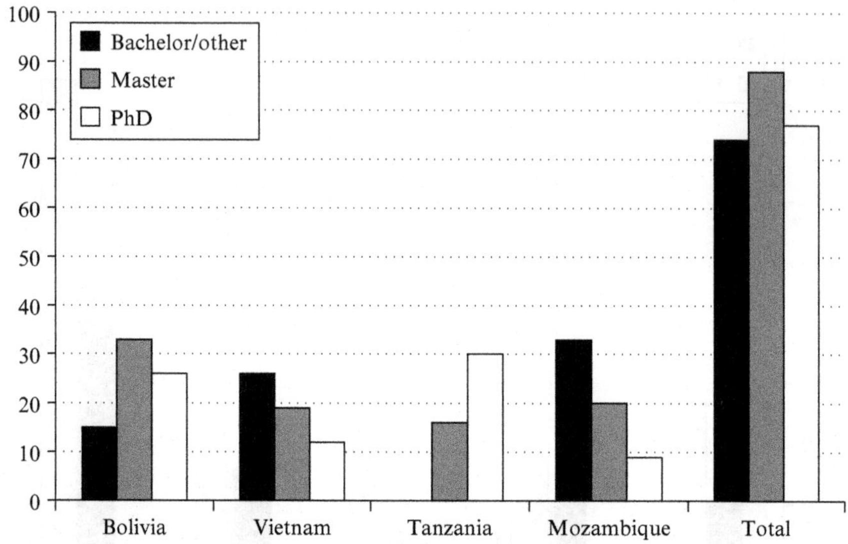

Source: National surveys.

Figure 8.3 Distribution of educational level by country in sample (number)

distributed between the three academic diploma of Bachelor's degree (31 per cent), Master's degree (37 per cent) and PhD (32 per cent) (Figure 8.3). The country samples differ somewhat: in Vietnam and Mozambique, the level of education among the respondents is lower than in the other two countries, with most researchers holding a Bachelor's or Master's degree. In Bolivia and Tanzania, PhD holders account for a larger share – 35 per cent and 61.2 per cent, respectively.

Table 8.2 summarizes the relative characteristics of the respondents. As discussed above, the samples from the four countries vary somewhat in terms of gender, age and educational level. Some of these variations are the result of differing country characteristics while others can be attributed to the realities of conducting surveys in developing countries with limited physical and electronic contact possibilities. Yet other variations stem from the limitations of the sample. However, it is our contention that, taken together, the opinions expressed in the survey provide a unique source of information for the analysis of the functionality of the national innovation systems in the case countries and a starting point for a discussion on how the research community and the policy sphere can better communicate and mutually reinforce the impact of research being carried out in developing

Table 8.2 Gender, age and education in sample (%)

		Bolivia	Vietnam	Tanzania	Mozambique
Gender	Women	28.0	36.8	15.2	50.8
	Men	72.0	63.2	84.8	49.2
Age	Under 30	4.1	29.9	15.2	15.3
	31–45	59.5	36.8	34.8	64.4
	Above 45	36.5	33.3	50.0	20.3
Educational level	Bachelor/other	20.3	45.6	0.0	53.2
	Master	44.6	33.3	34.8	32.3
	PhD	35.1	21.1	65.2	14.5

Source: National surveys.

countries – in other words, how to Close the Loop. In the following, the results of the national surveys are compared and analysed in a synthesizing discussion.

8.4 SOCIO-ECONOMIC RELEVANCE AND QUALITY OF RESEARCH

The general quality of research in developing countries is almost routinely perceived as inferior to work carried out by colleagues in the industrialized world. Research output indicators indeed support such a notion, with developing countries lagging behind in terms of number of academic publications, citation levels and patent rates. Although some of these countries, notably China, are catching up to the leaders, the general level of research as measured in research output indicators remains comparatively low.

Given this situation, it is striking that most of the researchers in the survey feel confident that the research carried out is both relevant and holds an adequate level of quality. Over 60 per cent of the respondents characterize the overall quality of research in the country as 'average', while 20 per cent maintain that the quality of the research is 'good' (Figure 8.4). For three of the case countries, as many as 88–97 per cent of the respondents described the quality of research as either 'average' or 'good'.

The dissenting country is Bolivia where a remarkable 44 per cent of the researchers labelled the research being performed in the country as 'bad'. Thus, while three of the case countries share a positive outlook, Bolivia has a considerably bleaker evaluation of the ability of the national innovation system to perform high-quality research. Part of the reason for such a

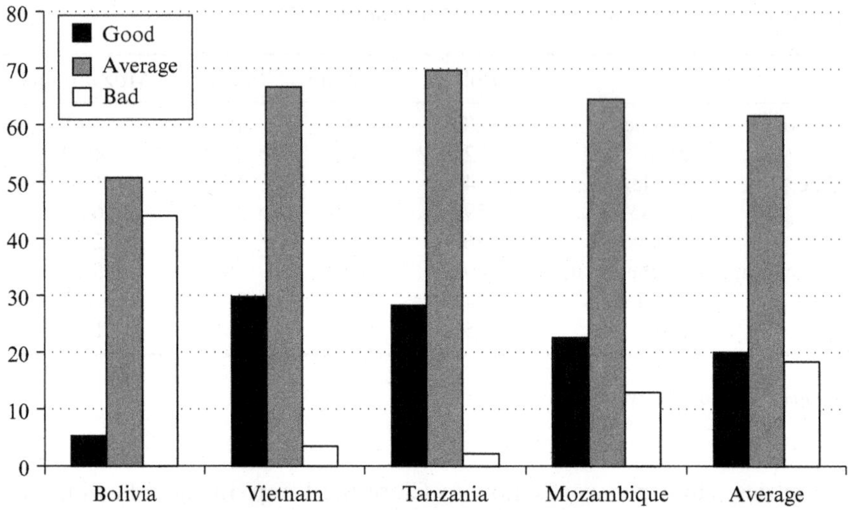

Source: National surveys.

Figure 8.4 What is your overall assessment of the quality of research in your country? (%)

discrepancy could be found in the fact that Bolivia is the country with the longest-running science and technology policy of the four. The sometimes ambitious but largely ineffective S&T policies introduced regularly in the country since the 1960s, and discussed at length in Chapter 3, could conceivably have created a certain fatigue and pessimism among researchers watching the continuous launching of new ineffectual STI plans struggling for space in the power sharing of government. As we shall see in the following, the national surveys substantiate a more optimistic outlook on the functioning of the innovation system in the other three countries, particularly Vietnam and Mozambique.

A further explanation, connected to the above, is the comparatively weak historic output of the STI sector in Bolivia despite its long existence. When we compare the output of the scientific community in terms of number of publications in the four countries in the Scopus database, we find that Bolivia has experienced the lowest growth over the period 1966 to 2013 (Figure 8.5). This is further accentuated when we employ the H Index measuring and combining scientific productivity (numbers of papers) and impact (citations) of individual researchers. Although Bolivia's 2014 H Index of 78 is higher than Mozambique's (66), it still trails Vietnam (133) and Tanzania (113) (SCImago, 2015).[2]

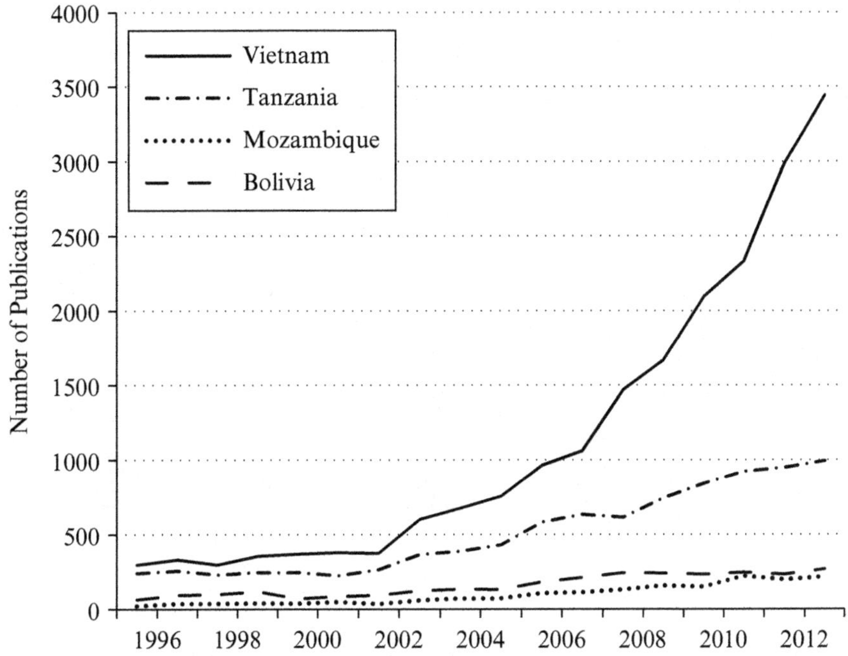

Source: SCImago (2015).

Figure 8.5 Number of scientific publication in the case countries, 1996–2013

A comment on the quality of research offered frequently by the surveyed researchers is that the problem is not the quality of research per se but that the results of the research are not put to use to the extent that they should. A common remark in the qualitative assessment is that the researchers are proud of the results but see systemic and linkage problems in the application and dissemination phase. They rather point to miscommunication between researchers and end-user as well as between researchers and policy makers than to the quality of the delivered work. When asked to what extent they feel that research results in their specific area are employed, the same pattern as for research quality can be observed (Figure 8.6). Most researchers perceived that the research output was 'satisfactorily' put to use (61 per cent), while only 28 per cent opined that research results were 'rarely' or never put to use.

Again, and consistent with the responses on the level of quality, Bolivia stands out as the country with the most negative view on whether or not

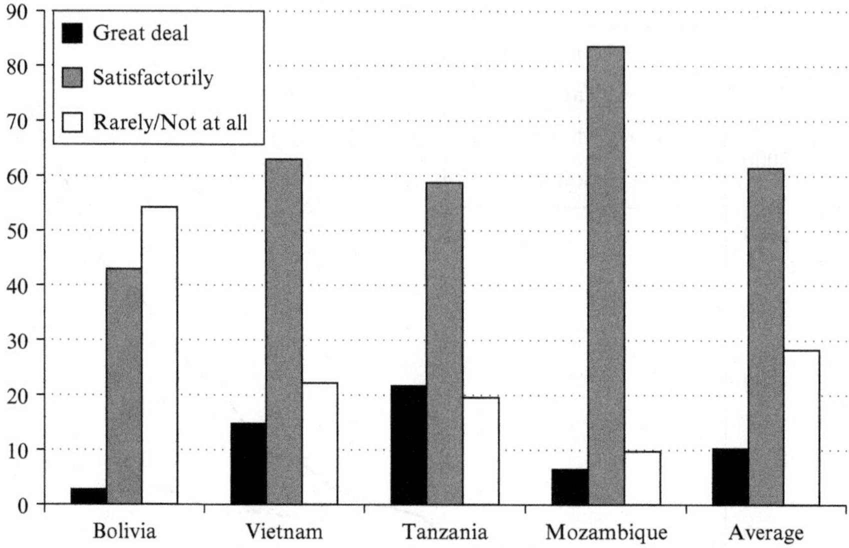

Source: National surveys.

Figure 8.6 To what extent do you think research results in your area are being put to use? (%)

research results are properly disseminated and used. A majority of the respondents in that country, 54 per cent, feel that research results are rarely or never put to use in the economy. Also, respondents in the other three countries are more negative than when asked about the quality of research; although the average is the same (61 per cent), researchers express a more pessimistic view on the use of research than on quality. This supports the previous assertion that it is failure in the application and dissemination phase rather than quality that hinders the efficient use of the fruits of research.

8.5 PROJECT SPECIFIC ISSUES

The freedom to formulate and pursue your own line of research is a cornerstone in the Humboldtian academic tradition. For today's researchers, however, the luxury of pursuing independent research has been impinged upon by increasing, and legitimate, demands for 'relevance' from funders, politicians, colleagues and society at large. The pressure for performing specific research tasks and adhering to external demands can be high when specific groups of users or funders determine the relevance of the research performed. Increasingly, calls for targeted research in the form of strategic and needs-driven research can limit the academic freedom of self-determination in research.

When asked about what influences the formulation of their research, the respondents evidently did not feel too constrained by internal and external pressures. On average, 60 per cent expressed that they felt free to follow their own research interest 'always or most of the time'. The responses are consistent between the case countries except for Vietnam where this figure was 45 per cent, probably a result of Vietnam having the youngest sample in the survey; the 'under 30' group experienced less opportunities than their older colleagues to follow their own research interest (Figure 8.7). Given the pessimistic view expressed by the Bolivian

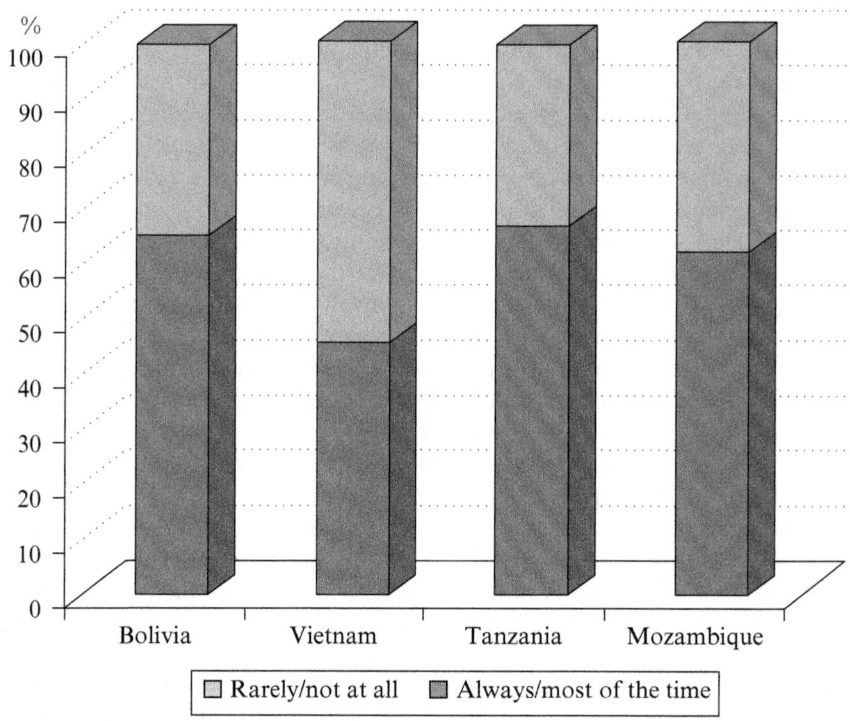

Source: National surveys.

Figure 8.7 How free are you to formulate research projects according to your own research interest? (%)

researchers on the low quality and underutilization of research results, it is also interesting to note that as many as 65 per cent of the respondents in Bolivia indicate that they do not experience any problems in following their own research interest. Clearly, the main problem for the S&T sector in that country does not emanate from a too tight governance structure or dirigisme but rather an inability of the sector to match demand for with supply of new knowledge.

The fairly positive perceived view on freedom of research notwithstanding, researchers always have to adhere to politico-academic realities in the quest for research funding. When asked about which factors are the most influential in research project formulation, complying with national development goals was the most frequent response, accounting for 29 per cent, closely followed by compliance with the priorities of funding agencies (27 per cent). Compliance with faculty or department research priorities scored lower (19 per cent) as did academic impact (18 per cent) (Figure 8.8).

This would seem to indicate a general willingness on behalf of the researchers to subordinate more narrow self-interests in academic positioning in favour of supporting national development goals. This runs contrary to the common belief that the publish-or-perish culture is increasingly driving researchers also in developing countries and that this creates strong incentives to publish in Western peer-reviewed journal rather than to produce local knowledge. Van Dalen and Henkens (2012) in a worldwide survey of demographers found that the pressure for publishing in that profession has resulted in a move away from policy and translating research outcomes for the public, towards publications intended for academic circles. They conclude that: 'a strong focus on academic publications tends to crowd out activities that may increase the amount of information available to policymakers and the general public' (van Dalen and Henkens, 2012, p. 1292). There are no reasons for assuming that this conclusion is not also valid in other researcher areas.

The relatively low score of the researchers in the case countries on the importance of academic impact in project formulation suggests either that academic pressure is not so prevalent in these countries or that the researchers have been able to combine academic pressure for publishing with the imperative of providing developmental impact. Either way, the general perception expressed by the researchers is that national developmental goals influence project formulation more than the academic incentive structure.

There are of course country differences. All countries score high on importance of national goals (over 20 per cent), with Mozambique reaching over 36 per cent, but in both Bolivia and Vietnam the importance

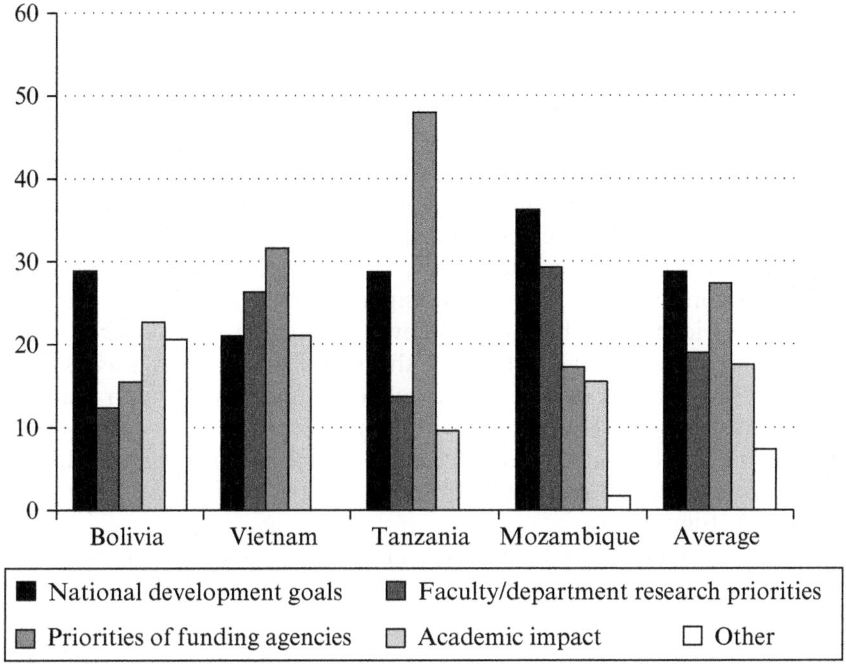

Source: National surveys.

Figure 8.8 Which of the following factors do you consider the most important when you formulate your research projects? (%)

of academic impact is also over 20 per cent, albeit lower than the importance of national development goals. Moreover, 7 per cent of the respondents reported the category 'other' as the most influential factor. When asked to specify, the most common explanation was that compliance with the other factors was secondary to the quality of the research proposal and that this factor was the most important factor guiding research formulation.

For the purpose of this book it is also interesting to note the strong influence of funding agencies. In Tanzania, a highly donor-dependent country in research, with almost 50 per cent of total research in the country supplied by donor organizations (Madikizela, 2013), 48 per cent of the respondents singled out complying with priorities of funding agencies as the single most important factor in project formulation. This begs the question of who determines the research agenda and to what extent research funding is aligned with the intentions of the policy as expressed in national development goals.

Closing the Loop hinges on good communication between researchers and policy makers in the process of policy formulation. Moreover, policies that are the result of this process are in turn interpreted and implemented by funding agencies and channelled to the research community in the form of funding priorities and calls for proposals. It is important that this operationalization of policy works as intended by the policy makers. In the survey, the researchers were asked about their opinion on the ability of funding agencies to align their expressed research priorities with national developmental goals. A large share of the respondents was critical of the way funding mechanisms were employed in support of developmental goals. The country with the most critical respondents was Bolivia with almost four out of five researchers expressing doubts on the functioning of the funding system, followed by Mozambique (52 per cent), Tanzania (46 per cent) and Vietnam (42 per cent) (Figure 8.9).

Such high numbers could indicate a mismatch between the interpretation

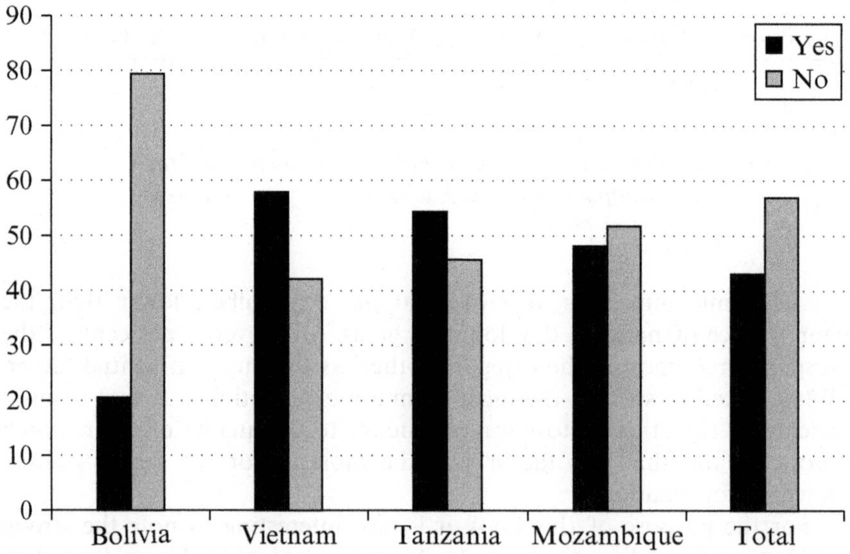

Source: National surveys.

Figure 8.9 *Do you think adequate measures are taken by the funding agencies to ensure alignment of research objectives with national developmental goals? (%)*

of national research priorities by funders and by researchers in all four countries as well as a vote of no confidence on the funding system by the researchers. The high degree of external influence in research funding by donor organizations and other international agencies has already been mentioned as an explanatory factor. Although donors in general strive to comply with and support national development objectives, the final selection of projects and emphasis on specific objectives undoubtedly is influenced by the overarching socio-economic and political goals of the donor. Researchers all over the world are quite adept in tweaking their project proposals to better fit the strategic goals of funding agencies. Moreover, the ambition of the donor community obviously is to be as effective and achieve as high an impact as possible, but as pointed out by an Organisation for Economic Co-operation and Development (OECD) study on the performance of funding for public research in higher education, evidence-based analysis of impacts is scarce also in industrialized countries and performance indicators are varied and constantly evolving (OECD, 2010). With such problems in evaluating results of their engagement in research support, it is perhaps understandable that donors may resort to following own priorities rather than strictly adhering to detailed national development priorities. In addition, end-of-the-project reporting to funding agencies tends to focus more on short-term quantitative output rather than on long-term outcomes relevant for national development goals.

The result of the survey is somewhat ambiguous on the influence of external research funding; while Vietnam with a comparatively low level of donor-funded research indeed is less critical to the functioning of the funding mechanisms, the negative perception of respondents in Bolivia is considerably higher than in the highly donor-dependent countries of Mozambique and Tanzania (Figure 8.9). A contributing factor to the case of Bolivia could be found in the professed non-existence of a properly functioning research funding system both for health and agriculture. Thus, many responses come from a feeling that in the absence of a funding system or mechanism it is difficult to judge policies or measures of the funding agencies.

In the responses we can also observe the only persistent gender difference of the survey. In all of the countries, women researchers are less critical of funding agency policies than their male colleagues. The biggest differences can be observed in Tanzania where only 14 per cent of the women, compared to over 51 per cent of the men, are critical of the funding system's ability to ensure the alignment of research objectives to national development priorities, and in Bolivia where male researchers are twice as likely to be critical as the women researchers. One reason for this could be that women tend to be more involved in research where national development policies coincide to a higher degree with donor policies, such as in health programmes and gender-specific research projects.

8.6 PERCEPTION OF IMPACT OF RESEARCH

Irrespective of the negative attitude towards the funding system, the respondents are generally quite confident that their research is meaningful and relevant. Over 84 per cent of the researchers state that they feel their research has had an impact on the local or national level in the country (Figure 8.10). This is fully consistent with the above finding of a positive assessment of the quality of the research system and also with the finding that research results in their respective area of expertise are applied to a 'satisfactory' level.

Researchers in all four countries expressed high rates of perception of the impact of research, with respondents in Vietnam unanimously replying yes to the question of whether their research is having any impact. Researchers in Bolivia reported a slightly lower level – although still high at 78 per cent – much in line with the finding that researchers in that

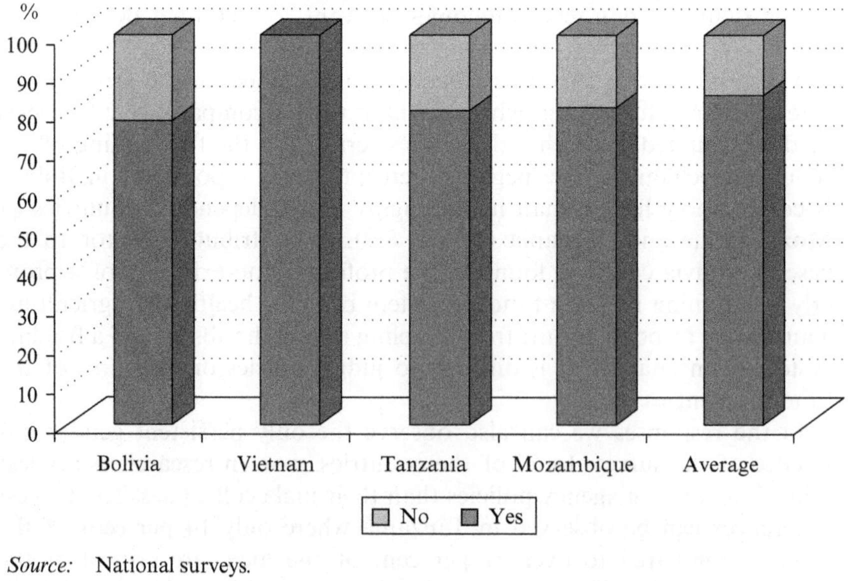

Source: National surveys.

Figure 8.10 Do you feel that your research results have had an impact on the local or national level?

country had the most negative view of the four countries on the quality of the research carried out. The high levels of perception of impact could to some extent be expected in this self-assessment of own research results, but the answers are consistent with opinions expressed in response to other questions.

The respondents that reported an impact of research in their own field were further asked to indicate what kind of impact of their research they have observed. With multiple answers possible – to allow for different kinds of impact in several projects – the respondents in all countries selected impact on social development as the category where they perceive their projects have had the most impact (Figure 8.11). Given the two scientific fields highlighted in the survey – agricultural science and medical science – it could perhaps be expected that the area where the effects of research would be the most prominent is the social lives of the end-users rather than support to industry through transfer of new technology. To a lesser extent, effects on economic development were also reported. Also here the focus on agriculture and health makes it seem likely that research results, such as new and/or more efficient medicine and vaccines as well as improved agricultural techniques and seed varieties, would result in better health or larger crops and translate into improved economic conditions.

Further, and particularly relevant for the theme of this book, researchers in both Bolivia and Vietnam reported limited impact on policy. In

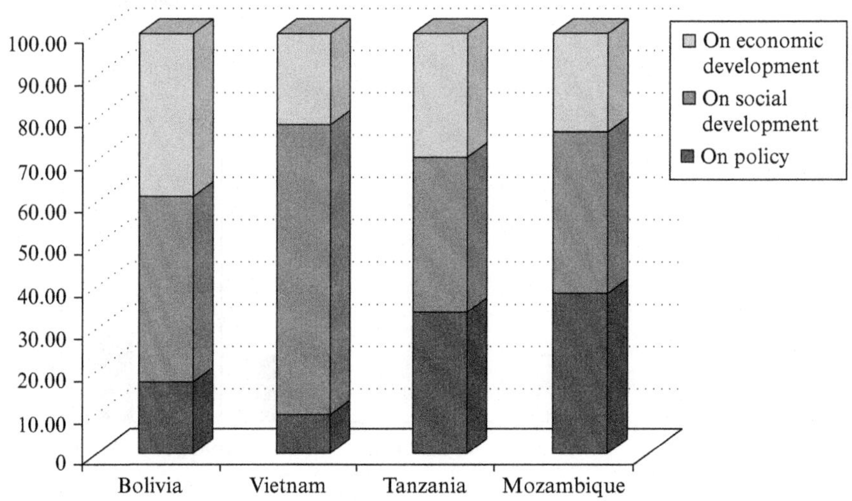

Source: National surveys.

Figure 8.11 What kind(s) of impact? (%)

Vietnam only 9 per cent of the respondents felt their research had any impact on policy. In Bolivia this percentage increases to 17 per cent. The other two case countries reported higher levels of impact on policy with 38 per cent in Mozambique and 34 per cent in Tanzania. These results indicate that many researchers, particularly in Vietnam and Bolivia, perceive that their research is disconnected from the policy formulation process and that their research does not inform policy makers in the priority setting to the extent it could.

The perception of being disregarded in policy formulation is obviously influenced by what the prospects are for researchers wanting to engage in a dialogue with policy makers or presenting policy-relevant and evidence-based research results in other forms. When asked what opportunities exist for researchers to present research results to policy makers and end-users, the same consistent pattern as for policy impact emerges. Many researchers in Vietnam (81 per cent) and Bolivia (39 per cent) reported that they felt that such opportunities did not even exist at all and in Mozambique this number was 27 per cent (Figure 8.12). The exception is Tanzania where none of the researchers felt that they never had any opportunity to present research results. In all countries except Vietnam, the majority stated that they at least sometimes had such opportunities.

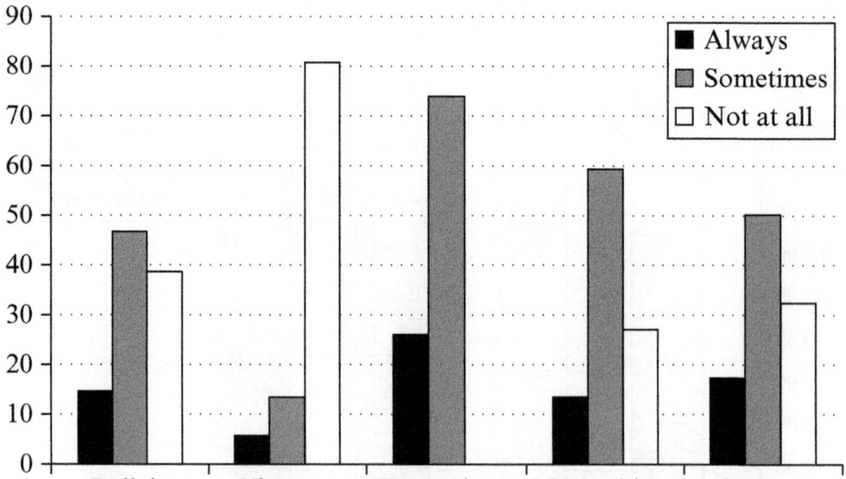

Source: National surveys.

Figure 8.12 Do you feel opportunities exist for researchers to present research results to policy makers and end-users? (%)

It is important to note that the responses did not only relate to policy makers but also to other types of end-users. This could to some extent explain the deviating responses from the respondents in Tanzania. In the qualitative responses several researchers expressed frustration with the non-communication with policy makers but were positive to the prospects of closely working with and introducing research results to the farming communities and citizenry that they serve.

While it is encouraging to note that half of the respondents feel that they sometimes have opportunities to present research results, it is striking that as many as 32 per cent state that they do not feel they have any such opportunities at all. Comments by the respondents indicate that it is in particular the possibility to interact with policy makers that is lacking. Few chances of engaging in a dialogue with policy makers have prompted some respondents to suggest the establishment of platforms for a constant and permanent collaboration between the research community and policy makers. These suggestions include establishing specific channels for translating and presenting research results to policy makers through regular seminars, workshops, roundtables, reports and briefing notes. But the respondents also recognize that Closing the Loop is a two-way street and that the research community needs to train researchers to be better in turning their research results into policy products, for example, how to write policy briefs, policy reports and organize policy dialogues. Such measures would go a long way in facilitating an improved understanding of the problems of communication and support a mutual learning process.

Others are less optimistic. In the chapter on Tanzania it is observed that policy makers often do not even have time to read policy briefs, much less full-length reports. In Bolivia it is pointed out that policy makers react more to political or social pressures than to scientific evidence. And in Mozambique one respondent lamented: 'is it really sensible to present scientific information to decision makers when we know it will not be used?'.

Irrespective of the ability of the research community and policy makers to improve their modes of interactions, the country cases have shown that the likelihood of research results being fed into the policy-making process increases when policy makers are involved or have an early interest in a specific project. Likewise, when research problems are defined and negotiated in partnership with potential users, the chances for use increase considerably. This points to the importance of early involvement of both policy makers and other end-users in the project formulation process.

8.7 CONCLUDING REMARKS

The survey set out to investigate to what extent the four case countries have been able to establish a national innovation system where research results are relevant for the developmental goals of the country as well as to inform policy making. How successful have they been? As discussed in Chapter 2, the four countries have made considerable progress in meeting the Millennium Developmental Goals. This would indicate that in particular Bolivia and Vietnam, with the best track record of the four countries, but also Tanzania and Mozambique have been able to gear at least part of their national innovation strategies towards combating pressing societal problems and promoting economic development.

The survey of the perception of researchers on the workings of the innovation system both confirms and contradicts this notion. It is clear that the respondents feel that their own research as well as the research of others in the country is both relevant and holds average to high quality. This, in spite of the lamentation in all countries that resources spent on research are much too low and not being put to efficient use. Rather than being an issue of quality, the researchers offer the view that the structure of the innovation system is not conducive to research results being put to use to the extent that they could. In other words, it is a system failure rather than a competence issue.

The exception to the overall positive view on quality and relevance of national research is Bolivia where there appears to be a severe miscommunication between knowledge producers and policy makers. In the opinion of the researchers, the rather linear research policy approach in the country hampers the ability of the innovation system to effectively address developmental problems – again a system failure. Despite the more pessimistic view on the overall quality of the research efforts, respondents in Bolivia – as well as in the other three countries – feel that their research indeed has a positive impact on development. Consistent with the critique of the functioning of the innovation system, a general belief among the researchers is that the impact would have been clearly better with an institutionalized and permanent dialogue between researchers and policy makers. Platforms for such interchanges of ideas are sorely missing in all four countries.

One prerequisite for Closing the Loop is that research carried out follows the priorities set out in national development policies. Therefore, it is important to note that the respondents express a willingness to follow national development priorities over more narrow academic demands in formulating and conducting research projects. The academic incentive system as well as conflicting or competing priorities set by donors can otherwise distort research efforts away from national priorities and towards more career-enhancing endeavours.

Perhaps surprisingly, gender is generally not a factor in the results of the survey. There is a slight tendency of women to be more positive in some of the responses but only in one case – attitudes towards funding agencies – is there a clear gender difference in all four countries with women expressing more positive attitudes towards the ability of the funding system to support national priorities. Similarly, no discernable persistent differences in response patterns could be observed when it comes to age and educational level.

In summary, all four countries are struggling to establish a virtuous cycle of mutual reinforcement between researchers and policy makers to inform policy with evidence-based research results.

Although still having a long way to go, the countries that would appear to have progressed furthest in this respect are Mozambique and Tanzania where the perception of the researchers is that they to a modest extent can rely on their research results reaching the intended end-users and on providing policy makers with input in the policy formulation process. For Bolivia and Vietnam the gulf between the research performers and the users of research results appears to be wider. However, if the survey presented in this book is any measure, the research communities in all four of the case countries are signalling a willingness to support a process towards Closing the Loop. Vital in this process of formulating and implementing sustainable development strategies is a capacity to plan for the future. Or, as one researcher in Mozambique succinctly puts it, 'I believe that the research policy should not continue to be so "populistic" and respond only to immediate problems. Research policy should be a look into the future, a contemplation on what kind of society we want to have in the future.' The ability of the national innovation system to do so will greatly influence the developmental outcomes in the four countries.

NOTES

1. We gratefully acknowledge financial support for the survey from the Swedish International Development Cooperation Agency (SIDA).
2. In comparison, developed countries tend to achieve a much higher H Index: USA leads with an index of 1648 followed by the UK with 1015, Germany 875 and France 811 (SCImago, 2015).

REFERENCES

Madikizela, M. (2013), *Mapping Research Systems in Developing Countries, Country Report: The Science and Technology System of Tanzania*, UNESCO, available at http://academic.sun.ac.za/unesco/data/Tanzania.pdf.

OECD (2010), *Performance-based Funding for Public Research in Tertiary Education Institutions*, Paris: OECD Publishing, available at http://dx.doi.org/10.1787/9789264094611-2-en (accessed 29 January 2016).

SCImago (2015), SJR – SCImago Journal & Country Rank, available at http://www.scimagojr.com (accessed 8 July 2015).

UIS (2015), UNESCO Institute of Statistics, Montreal.

van Dalen, H.P. and K. Henkens (2012), 'Intended and unintended consequences of a publish-or-perish culture: a worldwide survey', *Journal of the American Society for Information Science and Technology*, **63**(7), July, 1282–93.

Index

281